SYSTEMS AND SIGNAL PROCESSING OF CAPNOGRAPHY AS A DIAGNOSTIC TOOL FOR ASTHMA ASSESSMENT

SYSTEMS AND SIGNAL PROCESSING OF CAPNOGRAPHY AS A DIAGNOSTIC TOOL FOR ASTHMA ASSESSMENT

M.B. MALARVILI

Assoc. Prof., P. Eng., School of Biomedical Engineering and Health Sciences, Faculty of Engineering, Universiti Teknologi Malaysia, Malaysia

TEO AIK HOWE

Senior Consultant Emergency Physician, Emergency and Trauma Department, Penang General Hospital, Malaysia

SANTHERALEKA RAMANATHAN

Researcher, School of Biomedical Engineering and Health Sciences, Faculty of Engineering, Universiti Teknologi Malaysia, Malaysia

MUSHIKIWABEZA ALEXIE

Assistant Lecturer, University of Rwanda (UR), College of Science and Technology (CST)/Academic associate, Regional Center of Excellence in Biomedical Engineering and E-health (CEBE), Rwanda

Researcher at the School of Biomedical Engineering and Health Sciences at Universiti Teknologi Malaysia (UTM), Malaysia

OM PRAKASH SINGH

Researcher at the Group of Circuits and Systems, Tyndall National Institute, University College Cork, Cork, Ireland

ELSEVIER

ACADEMIC PRESS

An imprint of Elsevier

Academic Press is an imprint of Elsevier
125 London Wall, London EC2Y 5AS, United Kingdom
525 B Street, Suite 1650, San Diego, CA 92101, United States
50 Hampshire Street, 5th Floor, Cambridge, MA 02139, United States
The Boulevard, Langford Lane, Kidlington, Oxford OX5 1GB, United Kingdom

Notices
Knowledge and best practice in this field are constantly changing. As new research and
experience broaden our understanding, changes in research methods, professional
practices, or medical treatment may become necessary.

Practitioners and researchers must always rely on their own experience and knowledge in
evaluating and using any information, methods, compounds, or experiments described
herein. In using such information or methods they should be mindful of their own safety
and the safety of others, including parties for whom they have a professional
responsibility.

To the fullest extent of the law, neither the Publisher nor the authors, contributors, or
editors, assume any liability for any injury and/or damage to persons or property as a
matter of products liability, negligence or otherwise, or from any use or operation of any
methods, products, instructions, or ideas contained in the material herein.

ISBN: 978-0-323-85747-5

For information on all Academic Press publications visit our
website at https://www.elsevier.com/books-and-journals

Publisher: Mara Conner
Acquisitions Editor: Sonnini R. Yura
Editorial Project Manager: Isabella C. Silva
Production Project Manager: Prem Kumar Kaliamoorthi
Cover Designer: Vicky Pearson Esser

Working together
to grow libraries in
developing countries

www.elsevier.com • www.bookaid.org

Typeset by TNQ Technologies

Contents

Foreword

Asthma is a chronic inflammatory disorder that attacks the airways of the lungs. Asthmatic patients mostly experience persistent shortness of breath, wheezing, and chest tightness. Traditionally, asthma is detected based on physical assessment followed by pulmonary function test (namely spirometry or peak flow meter) to confirm the results. These tests are effort dependent. Capnography is a reliable alternative method for diagnosing asthma.

This book provides an overview of the respiratory system and its disorders, a substantial explanation of asthma, existing assessment methods, and limitations. Following that, this book delivers a detailed description of the design of interpretations of capnography as an alternative and advanced way for assessing and managing asthma. Finally, this book covers advances in the assessment of asthma with much focus on the currently developed system and signal processing methods using expired carbon dioxide (CO_2).

I congratulate the authors for applying their biomedical engineering knowledge to develop new systems and signal processing methods using expired carbon dioxide to provide updated information related to asthma and its new assessment methods. In addition, this book describes computational and transmission algorithms of capnography that would be the foundation for future developments. The extensive and rare information presented by the authors will also be beneficial to the researchers, academicians, students, and healthcare providers who take care of asthmatic patients.

Bhavani Shankar Kodali, MD
Professor of Anesthesiology
University of Maryland School of Medicine
Date: February 24, 2022

Preface

This book presents the advances in assessment of asthma with much focus on the currently developed system and signal processing methods, in which asthma is assessed and monitored using expired carbon dioxide (CO_2). This book is resourceful to researchers and academicians from biomedical engineering particularly in the area of biomedical instrumentation and signal processing. In addition, this book will be beneficial for healthcare professionals as it introduces a new approach in asthma diagnosis. Besides, it could serve as reference book for biomedical engineering students of both undergraduate and postgraduate level, who orient their careers in biomedical instrumentation. This book provides the overview of respiratory system and its disorders, the substantial explanation on asthma, the existing assessment methods, and their limitations. Following that, the book delivers a detailed explanation on the design of interpretations of capnography as an alternative and advanced method for assessment and management of asthma. Therefore, this book is a convenient reference to manufacturers of medical equipment, especially to get latest trend on asthma monitoring and emerging applications of capnography.

This book is divided into seven chapters and the corresponding references are provided at the end of each chapter. Chapter 1 provides the background of human respiratory mechanism. It includes a brief overview of respiratory system, symptoms of respiratory disorders and their effects, and a short introduction to asthma and its statistics. Chapter 2 further overviews on asthma and focuses mainly on the diagnosis and treatments of asthma. The current difficulties encountered by clinicians in identifying the potential solution for an asthmatic condition are discussed. Chapter 3 overviews the existing tools for the assessments of asthma namely spirometer, peak flow meter, and their limitations. Capnography as a promising alternative tool for asthma diagnosis is introduced in this chapter. Then, Chapter 4 further elaborates on the concept of capnography as CO_2 measurement tool, the phases of capnogram waveform, and its association for determining asthmatic condition of a patient. This chapter illustrates the significance of CO_2 features and available technology for the development of CO_2 measurement device namely mainstream and sidestream technologies. The role of capnography in determining the breathing state of a patient through the capnogram waveform is discussed in this chapter. Thereafter,

Chapter 5 discusses the quantification of capnogram waveform through signal processing techniques. The frequency and time domain features including various studies carried out on expired CO_2 signal are discussed in this chapter. Based on the type of features, the applied signal processing methods are discussed, to discriminate asthmatic and nonasthmatic conditions of a patient. Chapter 6 overviews the variety of CO_2 sensors available for development of a system for detecting exhaled breath which serves as a core unit for capnography. The advantages and limitation of each CO_2 sensors are elaborated in relation to enhancing capnography for detecting CO_2 from exhaled breath. Then, the last chapter of this book encloses the details of an approach executed in designing CO_2 measurement device, which serves as a simple capnography for use in diagnosis of asthma. Overall the system of CO_2 measurement device and the coherent explanations of its different components as well as the algorithms for features computation and transmission are presented in this chapter. Chapter 7 illustrates the available technology for capnography and its integration in the designed system to serve as an alternative approach and advancement for assessment and monitoring asthmatic conditions.

Acknowledgment

குறள்: ஒருமைக்கண் தான்கற்ற கல்வி ஒருவற்கு
எழுமையும் ஏமாப்பு உடைத்து.
விளக்கம்: ஒரு பிறவியில் தான் கற்ற கல்வியானது, ஒருவனுக்குத் தொடர்ந்து வரும் ஏழு
பிறப்புகளிலும் அவனைப் பாதுகாக்கும் சிறப்புடையது ஆகும்.

Kural: Orumaikkaṇ taaṉkaṟṟa kalvi oruvaṟku
eḻumaiyum eemaap uṭaittu.
Meaning: The learning, acquired by one in one birth, will continue to be a
fortress to one until the seventh birth.

As a matter of first importance, we would like to thank the Mercy of Lord for presenting His choicest grace and furnishing us with His ample of intelligence, perseverance, and constancy to complete the book.

We would like to express our respectful gratitude to the writers, technicians, students, colleagues, industry partners and research assistants who have provided the form of academic support, emotional, moral, financial, or while undertaking the challenge of completing this book.

Finally, we would like to express our warmest appreciation to our family, who have provided the infinite moral support and motivations. Thank you.

List of abbreviations

ABG	Arterial blood gases
AC	Alternating current
Arduino IDE	Arduino integrated development environment
CO_2	Carbon dioxide
COPD	Chronic obstructive parenchymal disease
CoV-19	Severe acute respiratory syndrome coronavirus two
CPR	Cardiopulmonary resuscitation
CT	Computer tomography
DC	Direct current
$EtCO_2$	End-tidal carbon dioxide
ETT	Endotracheal tube
FEV1	First forced expiratory volume
FFT	Fast Fourier Transform
FVC	Forced vital capacity
H1N1	Influenza virus
Heliox	Oxygen-helium mixture
ICSP	In-Circuit Serial Programming
IR	Infrared
LED	Light emitting diode
LRTIs	Lower respiratory tract infections
MAF	Moving average filter
MEMS	Microelectromechanical systems
MIR	Mid-infrared
NDIR	Nondispersive infrared
O_2	Oxygen
$PaCO_2$	Partial pressure of carbon dioxide
PCB	Programmable circuit boards
PEF	Peak Expiratory Flow
PEFR	Peak expiratory flow rate
$PetCO_2$	Peak end-tidal carbon dioxide
PSD	Power spectral density
RGM	Respiratory gas monitoring
RR	Respiratory rate
RTC	Real-time control
SABA	Short-acting bronchodilator
SARS	Severe acute respiratory syndrome
SD	Standard deviation
SD card	Secure digital card
SPI	Serial peripheral interface

SpO$_2$	Peripheral capillary oxygen saturation
SR	Slope ratio
TFT	Thin film transistor
URTI	Upper respiratory tract infections
USB	Universal serial bus
WHO	World Health Organization

List of figures

List of tables

CHAPTER ONE

The human respiratory system and overview of respiratory diseases

1.1 The primary role of the human respiratory system

Each human cell requires oxygen (O_2) and nutrients for its metabolism, which produces energy that is used by the cells to sustain cell activity. The main by-product of metabolism is carbon dioxide (CO_2), which dissolves into the small capillaries surrounding our tissues and is then brought by the circulating blood to be excreted from the body, mainly through the lungs (Fig. 1.1). The oxygen that is used comes from the air in the external environment. The respiratory system brings air from the external environment into the lungs with each inhaled breath. The inhaled air passes through the respiratory passages, to the alveoli in the lungs. The alveoli are the largest part of the lungs, comprising millions of small air-filled sacs with very thin walls, which are richly covered with capillaries. The proximity of the air-filled alveoli and the surrounding blood-filled capillaries ensure that O_2 can move easily from the alveoli into the blood to be carried to the rest of

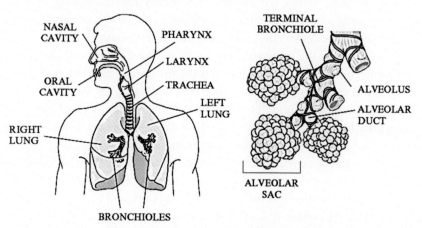

Figure 1.1 Diagram of the respiratory system illustrating the respiratory passageways. *Reproduced with permission from (Blanchard, 2005) Copyright 2005 Elsevier B.V.*

Systems and Signal Processing of Capnography as a Diagnostic Tool for Asthma Assessment
ISBN: 978-0-323-85747-5
https://doi.org/10.1016/B978-0-323-85747-5.00002-4

1

the body. On the other hand, CO_2 moves mainly in the opposite direction, brought by the circulating blood into the capillaries of the lungs, which surrounds the alveoli. Here, CO_2 diffuses into the alveoli, to be exhaled in the next breath.

1.2 The breathing process

During each inhaled breath, the muscles of the diaphragm and chest wall contract; the diaphragm moves downward, and the chest wall expands outward, resulting in an increase in the volume capacity of the chest cavity. This causes a net negative pressure within the chest cavity ("transpulmonary pressure"), which pulls air into, and expanding, the lungs. Through this mechanism, each breath draws in air from the exterior via the respiratory passages (upper airway, trachea, main bronchus, and smaller bronchioles) until it reaches the alveoli, which expands like millions of tiny balloons to receive that air (Yang et al., 2022). This is an active muscular process, i.e., muscles in the diaphragm and chest wall (often referred to as the muscles of inspiration) actively contract, expanding the chest cavity, to take in each breath. We are often not conscious of this muscular effort to take breaths. This is because these breaths are automatically and carefully regulated by the body allowing us to breathe without conscious effort. This is called quiet respiration, and this happens most of the time. However, these same regulating systems in our body also ensure that we are able to breathe faster and take deeper breaths when we need more O_2 or need to discard more CO_2, e.g., when we exercise, or after a breath-holding episode. This regulating system is carefully controlled by the body, maintaining the right amount of oxygen and carbon dioxide in the body.

Exhalation, on the other hand, is mainly a passive process. At the end of inhalation, the lungs are expanded, and the muscles of inspiration, the diaphragm, and muscles of the chest wall, have contracted. Exhalation is merely the relaxation of those muscles of inspiration. The elastic recoil of the structures of the lungs, together with the relaxation of the muscles at the end of inspiration, results in a net positive pressure in the chest, and expiration of air occurs through the same breathing passages. We can see that exhalation, unlike inhalation, requires no active muscle activity. Fig. 1.2 represents the changes in lung volume and pressure during respiration. The breathing cycle then repeats itself, which it does continuously throughout our life, from the very moment we take our first breath as a newborn, to that moment of our last at our death. A truly remarkable feat indeed. Similarly with inhalations,

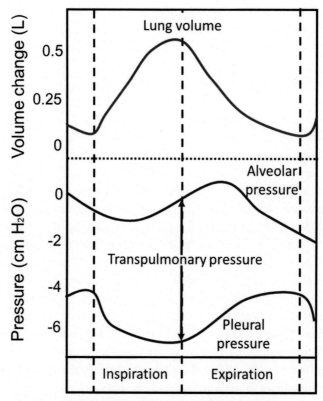

Figure 1.2 Changes in lung volume, and pressure differences during normal breathing.

exhalation during quiet respirations often occurs without our conscious effort or knowledge. We hardly ever need to remind ourselves to exhale (Harrison et al., 2021; Lukarski et al., 2022). However, sometimes we do need to use muscular activity to help exhalation, e.g., when we need to breath quickly during exercise, or in some respiratory diseases where exhalation may be impaired or obstructed.

The movement of air into the lungs occurs through the respiratory passages. The bronchioles are the ways for the flow of gas in and out of the air sacs. From the trachea they branch into left and right main bronchus, then each dividing into multiples of bronchioles. Bronchioles divide up to 16 times in total, until they are tiny in diameter. From the 16th to the 23rd division of the terminal bronchioles, they start to give rise to air sacs (alveoli) that combine to form the fragile soufflé that is the substance of our lungs. The diaphragm is almost entirely responsible for normal quiet breathing. The diaphragm contracts during inhalation, pulling downward and creating increased space within the chest cavity. This results in a net negative pressure, which pulls in the air. The diaphragm then relaxes during exhalation, allowing the elastic recoil of the lungs, chest wall, and abdominal tissues to compress the lungs resulting in a net positive pressure, which evacuates the air. However, during heavy respiration, the elastic forces are insufficient to induce the required rapid expiration, therefore extra effort is obtained mostly by abdominal muscle contraction, which pushes the abdominal contents upward against the bottom of the diaphragm, compressing the lungs (Gill et al., 2006). Raising the rib cage is the second way to expand the lungs. This is particularly used in active breathing when the rate and depth of breaths need to be increased. The ribs generally lie tilted downward in their natural resting posture. When the rib cage is raised, however, the ribs almost directly forward, causing the sternum to shift forward and away from the spine, increasing the anteroposterior thickness of the chest by nearly 20% during maximum inhalation compared to maximum exhalation (Elmedal Laursen et al., 2006; Masunaga et al., 2021). This mechanism, often referred to as a bucket handle mechanism, rapidly increases the anteroposterior diameter of the chest cavity, which causes a sudden and deep negative pressure, which draws in the breath quickly. Exhalation can occur by relaxation of those muscles during quiet breathing, which drops the ribs into their resting tilted downward position; or by contracting to opposing muscles that will depress the rib cage actively. Therefore, all the muscles that raise the chest cage are categorized as muscles of inhalation, and those muscles that depress the chest cage are known as muscles of exhalation.

1.2.1 Alveoli

At the microscopic level of the alveoli, every inhaled breath causes the expansion of each one of the millions of alveoli sacs in the lungs. Each inhalation brings new air, richer in O_2, into the alveoli. The oxygen quickly absorbs into the many surrounding capillaries and is then brought by these capillaries to the pulmonary veins and then to the heart and the rest of the body (Liu et al., 2011). During exhalation, the opposite happens; the air within the alveoli is now rich in CO_2, which had diffused out from the surrounding capillaries. As exhalation commences, this air is drained out of the alveoli bringing with it the CO_2 from the body. It is important to note that alveoli mostly do not empty out completely, but merely expand and contract with each breath, always with some residual volume left within it (Beltrán et al., 2022; Jiménez-Posada et al., 2021). The schematic representation of alveoli and its microscopic structure is shown in Fig. 1.3.

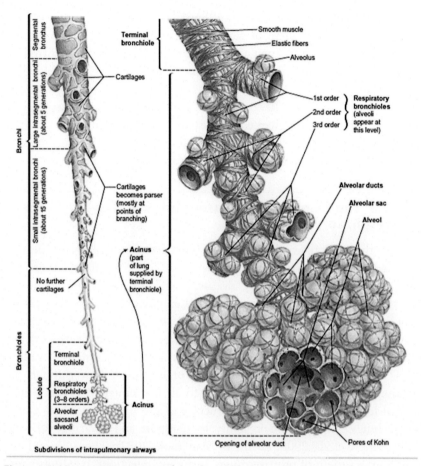

Figure 1.3 Microscopic structure of the alveoli. *Reproduced with permission from (Suarez et al., 2012) Copyright 2012 Elsevier B.V.*

In normal healthy lungs, each alveolus fills up uniformly in unison with the others; in response to the negative pressures created with inhalation. Similarly, during exhalation, the elastic recoil and relaxation of the muscles generally occur simultaneously, and each alveolus generally drains out uniformly with the others. Later in the chapter 2, we will see how this uniform movement of air in and out of the millions of alveoli simultaneously produces the standard shaped capnographic exhalation waveform.

However, there are in fact some subtle differences in alveolar filling, even in normal lungs. Due to gravity and the position of the body, parts of the upper lungs, bearing less weight of blood in the capillaries around it, have more air entering their alveoli. This is referred to as being "better ventilated." On the other hand, the lower lungs, which are more dependent, have a greater weight of blood in the capillaries around it, causing a relatively less air entering their alveoli, and relatively more blood surrounding the alveoli. This is referred to as being "better perfused" with blood. In short, in the lungs, the upper parts are generally better ventilated, poorer perfused as compared to the lower parts, which are better perfused, poorer ventilated. This is not a marked and usually not significant in the normal healthy lungs, but they become much more significant in lung diseases, which will be discussed in the subsequent parts of this chapter.

1.3 The respiratory diseases and how they affect the respiratory system

In the modern world, significant numbers of individuals are affected by some form of chronic lung disease, most common of which are the chronic obstructive pulmonary disease, which includes chronic poorly controlled asthma, chronic bronchitis, and emphysema (Witt et al., 2014). Respiratory diseases (including lung cancer) are responsible for a large and growing number of hospitalizations, ICU care for ventilator support and deaths. Respiratory diseases affect the human respiratory system in various ways. Airway diseases generally obstruct or impair air-flow through the breathing passages increasing resistance to air movement, whereas parenchymal lung diseases damage the structure of the bronchioles and alveoli of the lungs, impeding their ability to transport and exchange O_2 and CO_2 (Barrett et al., 2006; Levine & Marciniuk, 2022). These diseases can arise from a variety of primary causes, including infections, e.g., upper respiratory infections and pneumonias; acute or chronic inflammatory processes e.g., asthma, autoimmune lung disorders; structural damage from chronic exposure to irritants, e.g., chronic obstructive parenchymal disease (COPD) in smokers, occupational lung diseases; tumors e.g., cancers; or structural impediments to breathing or air exchange, e.g., pneumothorax, effusions, and pulmonary edema. The most common airway diseases are asthma and COPD. Asthma affects millions of individuals around the world in all age groups. Often starting as a disease of the young, in some it carries on throughout their adult life causing impairment to the daily activities and frequent visits to the hospitals. Despite advances in knowledge of the disease process leading to better control measures and treatments, asthma still claims thousands of lives each year, many at a relatively young age (Khaltaev & Axelrod, 2019). Equally, if not more devastating in terms of disease burden,

COPD is very much a disease of chronic smokers. Years of smoking cause a chronic inflammation in the airways, which essentially breakdown their structural and functional integrity. Occasionally, a new "novel" virus comes along, which nobody in the population has any innate immunity toward these novel viruses, e.g., SARS, H1N1, CoV-19 may cause a more severe disease affecting not just the upper respiratory tracts, but also the lower respiratory tracts and other parts of the body, where they can cause significantly more damage and cause a much worse form of illness resulting in deaths (Soriano et al., 2020). In these airway diseases, impairments to airflow through the respiratory passages are the main limitation bringing about the need for greater effort to breath and the symptom of breathlessness (World Health Organization, 2007).

In the developed world today, chronic lung diseases are a common cause of death. They are a primary cause of high healthcare costs due to multiple hospital admissions and costs of treatment. They result in productivity losses and days off-work, not to mention the adverse impact they have on the quality of life of those living with these diseases. But in the developing world, infections remain the principal cause of human deaths. In fact, respiratory infections have been the bane of mankind for aeons. The burdens carried along with the chronic lung diseases are shown in Fig. 1.4. Throughout the history of mankind, respiratory infections have been a significant cause of death at all ages, especially in younger children. To this day, respiratory infections remain the predominant cause of death in children and elderly population, especially in the developing world. By far the most common respiratory infections affect the upper respiratory tract (mouth, nose, and

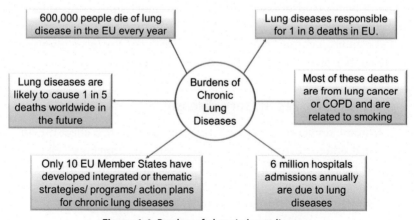

Figure 1.4 Burden of chronic lung diseases.

throat) and are commonly referred to as the upper respiratory tract infections (URTIs) (Korten et al., 2021). Each year, all of us usually encounter a bout or two of mild fevers with runny noses or sore throats. These are commonly viral infections, from a multitude of viruses that commonly cause URTIs. These viral infections tend to spread quickly with close contact, very rarely cause anything more serious that discomfort, and we often recover quickly on our own accord. The lower respiratory tract infections (LRTIs) affect the parts of the lungs that deal with O_2−CO_2 exchange, hence the greater danger that these types of infections can cause. LRTIs can be viral in origin, or they can come from a bacterial source (Brozek et al., 2015). Either way, the invading agent, and the body's response to it, gives rise to a series of changes in the lower respiratory tracts that result in more fluid and inflammation in the alveoli and terminal bronchioles. This impairs flow of air into and out or the alveoli; and as the alveoli become filled up with fluids hinders exchange O_2 and CO_2. The lack of O_2 and buildup of CO_2 will induce greater and greater respiratory efforts to overcome this life-threatening situation. If not adequately corrected, death may quickly ensue. Most respiratory diseases result in impairment of oxygen−carbon dioxide exchange, leading to a relative lack of O_2 and a build-up of CO_2 in the body. This immediately and automatically causes regulatory systems in the body to try to increase the rate and depth of breathing, in order to increase the amount of O_2 transported into, and CO_2 out of the lungs. The increased rate and depth of breathing result in the experience of being short-of-breath, of being breathless. This sensory experience, however, is subjective, and does not correlate to the degree of impairment of O_2−CO_2 exchange (Saers et al., 2021). For example, in some asthma patients, insensitivity to the feeling of being short of breath lead them to underestimate the severity of their asthma attack, thus delay seeking help and treatment. In many COPD patients, a chronic tolerance to low levels of O_2 exists without stimulating an increased respiratory effort.

1.4 Chronic respiratory diseases

Chronic respiratory diseases (CRDs) are illnesses that affect the lungs' airways and other structures. They have a strong inflammatory cell recruitment (neutrophil) and/or a destructive infection cycle (e.g., mediated by *Pseudomonas aeruginosa*). Asthma, chronic obstructive pulmonary disease, and acute respiratory distress syndrome are among the most frequent (Wang et al., 2021). CRDs are incurable; however, a variety of treatments that help dilate major airways and alleviate shortness of breath can help control symptoms and enhance the quality of life for those who suffer from them (Soriano et al., 2020). Respiratory diseases can be characterized in a variety of ways, including by the organ or tissue implicated, the type and pattern of accompanying signs and symptoms, and the disease's source (Basin et al., 2022). The section below elaborates the common respiratory disorders encountered in medical facility, emergency departments, and intensive care units.

1.4.1 Asthma and the other obstructive lung diseases

Asthma is a chronic inflammatory disorder of the airways, in which inflamed airways become hyper-responsive resultantly obstructing the airflow causing the patients to struggle for air. Airways are the pathways that carry the air in and out of the lungs. Asthmatic patients' airways, because of inflammation, produce thick, sticky discharge know as mucus (Gater et al., 2021). Furthermore, their airways also become tightened due to asthma attack. This inflammation, mucus and tightening of muscle cause the airways to be constricted as a result effecting the flow of air in and out of the lungs.

Asthma is not necessarily inherited, passed down to a child through their parents, but can also occur without any previous family medical history. One who is suffering from asthma his airways are far more sensitivity compared to a normal individual as a result, the airways get irritated and tightened easily. Things that cause irritation or tightening of airways are called triggers. These triggers can be caused by a variety of different things such as dust, allergies, smoke, cold, exercise and pollen, etc. The trigger varies from person to person, which causes asthma attack. Thus, it is very important for asthmatic patients to keep in mind their triggers and avoid things that can cause asthma attack. When asthma causes symptoms, it does so because of extensive narrowing of the bronchial tubes throughout both lungs and is likely very familiar with those symptoms such as shortness of breath, wheezing, cough, and tightness in the chest (Bagnasco et al., 2021). These muscles are involuntary muscles. It means that like the muscles that constrict the pupil of the eye in response to strong light or the muscles that move digested food through the intestines, they are not under the human direct or voluntary control. Under normal circumstances, when these muscles contract, the bronchial tubes that they surround narrow a little bit. Contraction of the bronchial muscles can occur over a period of minutes and can likewise reverse with relief of symptoms over a similar time frame (Yang et al., 2021). Although generally considered an acute intermittent disease, if uncontrolled over a period of years, severe chronic inflammation may cause permanent damage to the lung structures. Chronic debilitating lung diseases in children can result in deformity to the chest and rib cage becoming enlarged, causing a barrel chest, and both the functional residual capacity and lung residual volume become permanently increased (Patria & Esposito, 2013). The compensatory response of the body in response to inadequate O_2 and build-up of CO_2 is to increase the rate and depth of respirations (Brinkman & Sharma, 2018). This will improve O_2 and decrease CO_2 levels for a while but at the cost of needing more muscular activity and the constant unpleasantness of feeling short of breath. However, this is not an open-ended loop (Flatby et al., 2021; Kumar et al., 2019). There is a limit to how much increasing the rate and depth of respirations can work. In diseased lungs, merely increasing the rates and depths of respiration may be of little help if the air-exchange alveoli are severely impaired themselves. More importantly, there is a cost to the greater muscular activity needed to breath faster and deeper. All these muscles require oxygen to work and will generate a lot of CO_2 that needs to be expelled from the body (Shi et al., 2022). Ultimately, we will need more oxygen to work the muscles of respiration needed to get more oxygen. This is an unsustainable loop where the compensatory mechanisms are not be able to cater for the increased O_2 needs; if untreated this often results in the collapse and demise of the patient (Muiser et al., 2022).

1.4.2 The restrictive lung diseases

Restrictive lung diseases are a category of respiratory disease characterized by a loss of lung compliance, causing incomplete lung expansion and increased lung stiffness, such as in infants with respiratory distress syndrome (Distler et al., 2019). The diseases are defined based on the set of variances in the spirometry patterns. Spirometry evaluation indicates the changes in the lung volume, which is then classified as restrictive lung diseases. The common types of such conditions are lung expansion, reduced lung capacity, reduced lung volume, and reduced distensibility of lungs (Ramalho & Shah, 2021). The restrictive lung diseases are closely related to respiratory illnesses. This is because of the primary root cause of restrictive lung diseases, which may be originated from smokers, inhalation of toxic gases, and others. Hence, the disruptions occurred at respirator airways eventually cause damages to the lung, especially in the change of lung volumes and identified as restrictive lung diseases (Churg et al., 2010). The diseases encompassed of one-fifth of the overall lung diseases with mild restrictive syndromes. The primary cause of restrictive lung diseases is the disruption in lung parenchyma. It represents the intrinsic condition of lung infiltration due to toxins and inflammation. Besides intrinsic condition, extrinsic condition occurs with regard to the additional parenchymal conditions, which refers to the disruptions in the chest wall movements and neuromuscular functions. The restrictive lung disease is a branch of disrupted lung condition, which falls under one umbrella and deviates with the cellular infiltrates in a specified lung location. Debilitating fibrosis is a type of restrictive lung disease, which is commonly reported in emergency units and clinical conditions due to the self-controlled inflammatory conditions (Churg et al., 2010). The inflammatory movements develop changes in the alveolar interstitium and the peripheral bronchial structures. The changes in the alveolus may develop many severe lung conditions, such as hemorrhage, and edema. Cryptogenic organizing pneumonia, nonspecific interstitial pneumonia, sarcoidosis, idiopathic pulmonary fibrosis, systemic sclerosis, hypersensitivity pneumonitis, pulmonary Langerhans, and acute interstitial pneumonia are the intrinsic pulmonary parenchyma diseases, which lead to severe restrictive lung diseases (Won & Kryger, 2014; Wu et al., 2019).

1.4.3 The respiratory tract infections

Infections can affect any part of the respiratory system. They are traditionally divided into URT infections and LRT infections. URT infections are commonly identified with the intense irritation and swelling in the upper respiratory airways. Cough is the primary symptom of UTR infection as it involves the organ nose, sinuses, larynx, pharynx, and the airways. There are no possibilities to guess pneumonia from the symptoms of URT infections. Yet, the URT infection symptoms have been studied to diagnose early stage of pneumonia, although the patients have no history of any chronic diseases and genetic history of respiratory diseases (Zengel, 2019). Hence, the exact diagnosis of respiratory disease from URT infection symptoms is still crucial in medical field and the efforts on the improvements are taken. The most common upper respiratory tract infection is the common cold. However, infections of specific organs of the upper respiratory tract such as sinusitis, tonsillitis, otitis media, pharyngitis, and laryngitis are also considered URT infections (Bergmann et al., 2019). The primary cause of URT infection is the presence of viruses and bacteria in the upper respiratory airways. Due to the similar symptoms for every URT infections, a variety of individual test have been conducted to detect a specific type of virus or bacterial infection. The usual type of preliminary test for URI patients is performed to screen for common cold, acute bronchitis, respiratory distress, and influenzas. As the respiratory infections are caused by bacteria and virus, it is highly infective to others. The common cold and sneezing in a group of people have high risk of respiratory infections to get infected (Calapodopu-los et al., 2021; Ruscic et al., 2017). Hence, people with URT infections are advised to stay away from crowd and have no close contact with children, which can increase the risk of URT infections. Besides, the inhalation of cigarettes by smokes and passive smokers tends to show severe symptoms of URT infection and highly capable of infective a large group of people, due to the high multiplying rate of virus and bacteria. The infection happens with the direct invasion of infected upper respiratory mucosa from one to another organism. In most situation, the invasion occurs in the form of liquid in the air or by touch (David & Cunningham, 2019). Human noses consist of hair lining, mucus, and angles between nose and pharynx act as the barrier for preventing the invasion of infected mucous. Being the predominant symptom of URT infection, common cold is usually caused by microbes such as adenovirus, enterovirus, respiratory syncytial virus, parainfluenza vi-rus, rhinovirus, and coronavirus. Among these, rhinovirus is the most com-mon pathogen infecting the URT and results in almost 80% of all kinds of URT infections (Hwang et al., 2021; Ziou et al., 2022). Rhinovirus reaches the anterior nasal mucosa and rapidly replicates, as the cold symptoms appear in 10–12 h of virus infection. The common cold may persist up to 10 days with medication, as the pathogen infection results in rhinorrhea and nasal obstruction, which causes continuous mucus production and sneezing (Bianco et al., 2022; Xu et al., 2021). About 45% of respiratory disorders are caused by LRT infection, which also increases the mortality rate due to respiratory illnesses. In comparison to URT diseases, LRT infections require rapid and early diagnosis to prevent morbidity, lower the hospitali-zation risk, and ensure simple therapeutics. With the advancement of current medical facilities, LRT-based diseases can be detected at early stage with frequently empirical treatment (Gentilotti et al., 2022; Waghmare et al., 2019). Some bacteria with nonvirulence form a flora from upper to lower respiratory tract. The acute exacerbations of the bacteria infection cause mild bronchitis. The LRT disrupted with the interaction between respira-tory epithelial cell and bacterial cell. In general, and mild condition, the host defense mechanism removed the bacteria colonization through mucous clearance. However, the bacteria infection may disperse into the pooled mucous, which cause huge spread of infection and damage the respiratory epithelial cells (Almangour et al., 2021; Vardakas et al., 2018).

1.4.4 The diseases affecting the pleural cavity

Pleural cavity is a form of respiratory and pulmonary disease originated from viral infection. The virus presence at the pleura of lung walls causes disruptions to the thin space and layers of pleura. A collection of fluid in the pleural cavity is known as a pleural effusion (Campion et al., 2021). This may be due to fluid shifting from the bloodstream into the pleural cavity due to conditions such as congestive heart failure and cirrhosis. It may also result from the inflammation of pleura itself as it can occur with infection, pulmonary embolus, tuberculosis, mesothelioma, and other conditions. A pneumothorax is a hole in the pleura covering the lung allowing air in the lung to escape into the pleural cavity (Bara et al., 2021). The affected lung "collapses" like a deflated balloon. A tension pneumothorax is a particularly severe form of this condition where the air in the pleural cavity cannot escape, so the pneumothorax keeps getting bigger until it compresses the heart and blood vessels, leading to a life-threatening situation (Sorino et al., 2022).

1.4.5 The pulmonary vascular diseases

Conditions that impact the pulmonary circulation are known as pulmonary vascular disorders. A blood clot that forms in a vein, breaks free, travels through the heart, and lodges in the lungs is known as a pulmonary embolism (thromboembolism). Large pulmonary emboli can be devastating, resulting in death in a matter of minutes. Fat embolism (especially after bony injury), amniotic fluid embolism (with complications of labor and delivery), and air embolism (iatrogenic—caused by invasive medical procedures) are a few other substances that can embolize (travel through the blood stream) to the lungs, but they are much more rare (Llucià–Valldeperas et al., 2021). Hypertension of the pulmonary arteries, also known as pulmonary arterial hypertension. It is most usually idiopathic (meaning it has no known cause), although it can also be caused by the effects of another disease, especially COPD. Core pulmonale is a condition that causes strain on the right side of the heart. Pulmonary edema is a condition in which fluid leaks from the capillaries of the lungs into the alveoli (or air spaces) (Ghigna & Dorfmüller, 2019; Lu et al., 2019). Pulmonary edema is a condition in which fluid leaks from the capillaries of the lungs into the alveoli (or air spaces). Congestive heart failure is the most common cause. Blood spilling into the alveoli due to pulmonary bleeding, inflammation, and damage to capillaries in the lung. Blood may be coughed up because of this. Auto-immune illnesses such as granulomatosis with polyangiitis and Goodpasture's syndrome can cause pulmonary bleeding (Levine et al., 2019).

1.5 The common symptoms of respiratory diseases

Respiratory diseases run the gamut from mild to life-threatening, from acute intermittent attacks to chronic life-long illnesses. Respiratory diseases are identified based on the presence of certain respiratory symptoms e.g., cough, discomfort in the upper respiratory passages, sputum, shortness of breath, or difficulty in breathing. Cough serves as a protective reflex, an expulsive effort, which clears the upper airways of any obstructions or secretions. Cough usually signifies irritation of the airway passages commonly seen in URTIs, or presence of sputum due to increased secretions or mucous, which need to be expelled to keep the breathing passages clear. Large quantities of sputum produced from the respiratory passages are a common cause of cough; sputum that is discolored, purulent or foul-smelling suggests pneumonia or other infections of the LRTs (Maio et al., 2019). Patients may also complain of hoarseness of voice, discomfort during swallowing, which suggests irritation at the larynx and throat. Noisy breathing is an important symptom pointing toward obstructions to the airway passages, which cause a whistle-like effect producing polyphonic high-pitched sounds of wheeze in asthma. Sometimes upper airway obstruction produces lower pitched sounds of stridor, which should alert the healthcare provider to the possibility of a medical emergency. Painful breathing is relatively uncommon and usually suggests injury or impairment to the structures of the chest wall or the pleural space (Staso et al., 2021). Other symptoms such as fever suggest presence of infection, pain may suggest areas of inflammation and mechanisms of injury point to areas injured. Some symptoms point toward a more serious disease, especially symptoms of shortness of breath or difficulty in breathing. When this exacerbates into life-threatening conditions, failure of other organs is often noted with their concomitant findings—altered mental state indicates impairment of the neurological system, hypotension indicates circulatory collapse and worsening vital signs warn of shock and impending collapse. Table 1.1 indicates the differential diagnoses of shortness of breath, in regard with acute and chronic respiratory illness.

Rapid breathing rates and deeper breaths may be the only sign seen in early respiratory disease. This is part of the compensatory mechanisms that the body tries to incorporate in response to worsening O_2 and CO_2 levels

Table 1.1 Differential diagnoses of shortness of breath.

Acute causes	Chronic causes
Respiratory	Respiratory
Asthma	COPD
Acute exacerbation of COPD	Pulmonary fibrosis
(infective or noninfective)	Pleural effusion
Lower respiratory tract infection	Bronchiectasis
Pulmonary embolism	Cystic fibrosis
Pneumothorax	Chronic infection/tuberculosis
Cardiac	Cardiac
Acute pulmonary edema	Heart failure (chronic)
Decompensated heart failure	Chronic arrhythmias
Other	Others
Panic attacks/Psychosomatic	Anaemias
Metabolic acidosis	

Adapted from Oxford Medical Education.

in the blood. If this compensatory mechanism works to correct the O_2 and CO_2 levels, patients may often be unawares and then none is the wiser. But if the disease is worsening, breathing rates increase and the efforts of breathing become more labored, patients will start to complain of difficulty of breathing (Erdoğan et al., 2021; Hassan Lemjabbar-Alaouia et al., 2017). The conscious awareness of the inadequacy of breaths and the conscious need to take the next breath quickly is a highly stressful situation and causes significant distress to the patient. The physical signs of respiratory impairment at this stage are often exemplified by the use of accessory respiratory muscles to aid breathing efforts. As a quick reminder, during quiet respiration, only the diaphragm and some of the chest wall muscles are performing the work of normal breathing. However, when greater and greater effort is required, more and more muscles are called into action to support the respiratory effort (Calle Rubio et al., 2021). These muscles in the neck and chest wall are called the accessory muscles of breathing, and they are only used to aid breathing when extra effort is needed. When clinicians detect that these accessory muscles of breathing are in use, it signifies the diagnosis of respiratory impairment. Some postural positions assumed by breathless patients to support their breathing efforts are useful signs also pointing to significant respiratory impairment especially in younger children (Syed et al., 2021). They are common referred to as the tripod position seen in children with respiratory difficulties. Accessory muscle use and tripod positioning are markers of severe respiratory impairment. Fig. 1.5 shows the common causes result in respiratory difficulties in children.

Clinical findings may be identified during physical examination by doctors. These may include added sounds to breathing like wheezes and stridor (which have been explained above), or snoring, which suggests partial intermittent upper airway obstruction. Grunting, another added sound during breathing, is audible at the end of expiration when a forced positive pressure "grunt" is made to prevent collapse of the threatened alveoli. This is also a sign of severe alveolar disease, where there are significant and widespread alveolar collapses. Percussion of the chest may show areas of hyperresonance, which suggests pneumothorax, or dullness, which may indicate collapse or effusions. Auscultation may reveal added sounds such as crackles, which may point out pneumonias or pulmonary edema.

1.6 The breathless patient

It should be clear by now that the breathless patient is the most common presentation of serious respiratory diseases. Having said that, we cannot

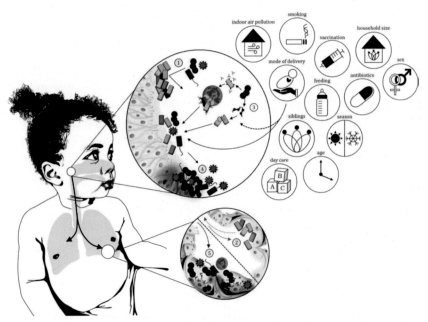

Figure 1.5 Sources of respiratory illness in children. *Reproduced with permission from (Claassen-Weitz et al., 2021) Copyright 2021 Elsevier B.V.*

ascertain that all breathless patients have underlying respiratory disease, for breathlessness is also associated with various types of cardiac, renal, muscular, and psychological disorders. *All that is breathless is not merely in the lungs.* This has been known for centuries; and it has for this long time that doctors have grappled with the diagnosis of the breathless patient. In fact, the actual Greek word asthma was named by Hippocrates, the famed physician, derived from the verb *aazein* meaning to exhale with an open mouth or pant (Marketos & Eftychiades, 1986). Even so, it is unclear whether Hippocrates meant asthma the clinical entity as we know it now, or the symptom, which is shared with other diseases. It was the later work of Aretaeus of Cappadocia and Galen, who better defined the features of asthma. Ancient physicians have noted that a multitude of diseases cause breathlessness. Apart from diseases of the lungs, they would have noted breathlessness complaints in patients with diseases of the heart, internal organs, infections of all kinds, imbalances of the humor and internal vapors, etc. It was a most exasperating diagnosis to consider for the possibilities were wide and varied. The limited knowledge of physiology and the reliance on ancient beliefs just made accurate diagnosis much more difficult. More astute physicians began to notice that certain symptoms would group together commonly (Sharma et al., 2022). For example, intermittent wheeze in younger children, which totally disappeared, only to recur again and again was diagnosed as asthma. Elderly patients who had breathlessness and swelling of the lower limbs were more of a problem of the heart. Fevers would suggest infections. Caseating lymph nodes proved tuberculosis as the cause. Even then, some presentations were just beyond the understanding of the day. For example, in some families, children would die in their youth, with fever, and a fruity odor in their breath, as they panted away until their last breath (Dreschfeld, 1886). We now know this to be Diabetic Ketoacidosis, a common complication of Type 1 Diabetes Mellitus; but before the discovery of insulin in the early 20th century, this was a dreaded diagnosis with almost universal mortality in sufferers with the genetic predisposition toward this disease.

1.7 Understanding the pathophysiology of breathlessness

Today, we have a much more detailed understanding of the disease processes of many conditions that cause breathlessness. Clinicians are more able to diagnose and differentiate the various presentations of breathlessness to come to a more accurate diagnosis and decide a specific treatment plan.

However, as our population lives to an older age, and encounters more medical illnesses, patients today often have more than one disease at the same time (Sung et al., 2020). Clinicians today have the distinctly more difficult and challenging task to try to identify the main disease causing the breathlessness among the several medical illnesses that the patient has. Therefore, various diagnostic tools and monitoring modes are required to aid clinicians in recognizing respiratory diseases better and to differentiate them from the other nonrespiratory diseases.

Various diagnostic tools and monitoring modes of respiratory diseases identify specific areas affecting the respiratory system. Structural changes within the respiratory system are often diagnosed using chest X-rays, ultrasound examinations or CT scans. These images are often diagnostic in itself; but with the advent of artificial intelligence, the accuracy of some models is already outperforming many qualified radiologists. More invasive procedures such as bronchoscopy or CT-guided biopsies may be needed to obtain samples for confirmatory biopsy tests, which will identify the nature of tumors, infections, or autoimmune diseases (Iizuka et al., 2019).

Further elucidation of specific respiratory diseases usually requires functional testing. Lung function tests are able to determine specific functional limitations of the lungs beyond the structural limitations identified predominantly by imaging techniques. Obstructions to airflow are identified by flowmeters and spirometry tests. In these tests, patients actively and forcefully blow into devices to measure the highest flow rate that they can generate. Flow problems, which are commonly seen in asthma and COPD, can be rapidly diagnosed with such tests, if they are able to perform them. Measurement of the capacities of the lungs indicates to doctors whether the underlying disease causes more of a restrictive inhibition of the lungs, or an obstructive impairment of airflow. These tests give doctors a better idea of the type of functional abnormality affecting the patient's respiratory system (Marshall et al., 2021). Fig. 1.6 illustrates the lung capacity measurement parameters, in relation to the total lung capacity. However, the main problem with lung functions tests is that they require active and forceful voluntary breaths performed by the patient. Often, as a result of their breathlessness itself, patients are not able to perform these forceful exhalations well enough to give an accurate depiction of their functional status of their lungs. Abnormalities in oxygen—carbon dioxide exchange can be measured by the levels of O_2 in the blood using pulse oximetry, or more accurately by blood gas measurements of levels of O_2 and CO_2 in the blood. This would indicate the final common path in the movement of O_2 and

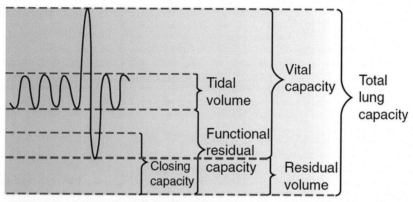

Figure 1.6 Lung capacities measures in lung function tests. *Reproduced with permission from (Culver, 2012) Copyright 2012 Elsevier B.V.*

CO_2 to and from the external environment and the cells of the body. There is unfortunately no single ideal diagnostic tool or monitoring mode because many different respiratory diseases overlap in their presentations, or may cause similar abnormalities between them; in addition, some functional tests require active efforts by the patient, which they are not able to perform especially when they are acutely breathless (Lewthwaite et al., 2021). The medical world therefore continues to search for better diagnostic and monitoring tools for common respiratory diseases. To understand this better, in the next chapter, we shall look in more detail at asthma, at how it is diagnosed and treated; and the difficulties facing clinicians in making the right decision for their patients. And a potential solution to this problem.

References

Almangour, T. A., Garcia, E., Zhou, Q., Forrest, A., Kaye, K. S., Li, J., Velkov, T., & Rao, G. G. (2021). Polymyxins for the treatment of lower respiratory tract infections: Lessons learned from the integration of clinical pharmacokinetic studies and clinical outcomes. *International Journal of Antimicrobial Agents, 57*(6), 106328. https://doi.org/10.1016/j.ijantimicag.2021.106328

Bagnasco, D., Povero, M., Pradelli, L., Brussino, L., Rolla, G., Caminati, M., Menzella, F., Heffler, E., Canonica, G. W., Paggiaro, P., Senna, G., Milanese, M., Lombardi, C., Bucca, C., Manfredi, A., Canevari, R. F., Passalacqua, G., Guarnieri, G., Patella, V., … Lo Cicero, S. (2021). Economic impact of mepolizumab in uncontrolled severe eosinophilic asthma, in real life. *World Allergy Organization Journal, 14*(2). https://doi.org/10.1016/j.waojou.2021.100509

Bara, A., AlKhatib, M. O., & Alkhayer, I. (2021). A rare case of numerous hydatid cysts in the pleural cavity without extrapleural involvement. *Annals of Medicine and Surgery, 65*(April), 102290. https://doi.org/10.1016/j.amsu.2021.102290

Barrett, M. L., Smith, M. W., Elixhauser, A., Honigman, L. S., & Pines, J. M. (2006). Utilization of intensive care services, 2011: Statistical brief #185. Healthcare cost and utilization project (HCUP). *Statistical Briefs, 15*(6), 1–14. http://www.ncbi.nlm.nih. gov/pubmed/25654157

Basin, S., Valentin, S., Maurac, A., Poussel, M., Pequignot, B., Brindel, A., Poupet, G., Robert, C., Baumann, C., Luc, A., Soler, J., Chabot, F., & Chaouat, A. (2022). Progression to a severe form of COVID-19 among patients with chronic respiratory diseases. *Respiratory Medicine and Research, 81.* https://doi.org/10.1016/j.resmer.2021.100880

Beltrán, G., Navajas, D., & García-Aznar, J. M. (2022). Mechanical modeling of lung alveoli: From macroscopic behaviour to cell mechano-sensing at microscopic level. *Journal of the Mechanical Behavior of Biomedical Materials, 126.* https://doi.org/10.1016/j.jmbbm.2021. 105043

Bergmann, M., Ballin, A., Schulz, B., Dörfelt, R., & Hartmann, K. (2019). Therapie des akuten viralen Katzenschnupfens. *Tierarztl Prax Ausg K Kleintiere Heimtiere, 47*(2), 98–109 [Treatment of acute viral feline upper respiratory tract infections].

Bianco, A., Licata, F., Nobile, C. G., Napolitano, F., & Pavia, M. (2022). Pattern and appropriateness of antibiotic prescriptions for upper respiratory tract infections in primary care paediatric patients. *International Journal of Antimicrobial Agents, 59*(1). https://doi.org/ 10.1016/j.ijantimicag.2021.106469

Blanchard, S. (2005). Anatomy and physiology. In *Introduction to biomedical engineering* (2nd ed., pp. 73–125). Elsevier Inc. https://doi.org/10.1016/B978-0-12-238662-6.50005-7

Brinkman, J. E., & Sharma, S. (2018). *Physiology, respiratory drive.* StatPearls. http://www. ncbi.nlm.nih.gov/pubmed/29494021

Brozek, G., Lawson, J., Szumilas, D., & Zejda, J. (2015). Increasing prevalence of asthma, respiratory symptoms, and allergic diseases: Four repeated surveys from 1993-2014. *Respiratory Medicine, 109*(8), 982–990. https://doi.org/10.1016/j.rmed.2015.05.010

Calapodopulos, N. V. I., Sawan-Mendonça, M. M., da Silva, M. V., Oliveira, C. J. F., Weffort, V. R., Rodrigues, D. B. R., & Rodrigues, V. (2021). Association of recurrent upper respiratory tract infections with low production of oxygen intermediates in children. *Jornal de Pediatria, 000*(xxx), 1–7. https://doi.org/10.1016/j.jped.2021.09.008

Calle Rubio, M., Rodríguez Hermosa, J. L., Miravitlles, M., & López-Campos, J. L. (2021). Knowledge of chronic obstructive pulmonary disease, presence of chronic respiratory symptoms and use of spirometry among the Spanish population: CONOCEPOC 2019 study. *Archivos de Bronconeumologia, 57,* 741–749. https://doi.org/10.1016/j.arbr.2021. 10.003

Campion, E., Mallah, S. I., Azhar, M., O'Keeffe, D., & Hameed, A. (2021). A multidirectional two-tube method for chemical pleurodesis could improve distribution of the sclerosing agent within the pleural cavity — a pilot study. *Annals of Medicine and Surgery, 68*(May), 102697. https://doi.org/10.1016/j.amsu.2021.102697

Churg, A., Müller, N. L., & Wright, J. L. (2010). Respiratory bronchiolitis/interstitial lung disease: Fibrosis, pulmonary function, and evolving concepts. *Archives of Pathology & Laboratory Medicine, 134*(1), 27–32. https://doi.org/10.5858/134.1.27

Claassen-Weitz, S., Lim, K. Y., Mullally, C., Zar, H. J., & Nicol, M. P. (2021). The association between bacteria colonizing the upper respiratory tract and lower respiratory tract infection in young children: A systematic review and meta-analysis. *Clinical microbiology and infection.* https://doi.org/10.1016/j.cmi.2021.05.034

Culver, B. H. (2012). Chapter 9 - pulmonary function testing A2 - spiro, stephen G. In *Clinical respiratory medicine* (4th ed., pp. 133–142) https://www.sciencedirect.com/science/ article/pii/B978145570792800009X

David, S., & Cunningham, R. (2019). Echinacea for the prevention and treatment of upper respiratory tract infections: A systematic review and meta-analysis. *Complementary Therapies in Medicine, 44*(February), 18–26. https://doi.org/10.1016/j.ctim.2019.03.011

Distler, O., Volkmann, E. R., Hoffmann-Vold, A. M., & Maher, T. M. (2019). Current and future perspectives on management of systemic sclerosis-associated interstitial lung disease. *Expert Review of Clinical Immunology, 15*(10), 1009–1017. https://doi.org/10.1080/ 1744666X.2020.1668269

Dreschfeld, J. (1886). The bradshawe lecture on diabetic coma. *British Medical Journal,* *2*(1338), 358—363. https://doi.org/10.1136/bmj.2.1338.358

Elmedal Laursen, B., Mulvany, M. J., & Simonsen, U. (2006). Involvement of guanylyl cyclase, protein kinase A and Na+K + ATPase in relaxations of bovine isolated bronchioles induced by GEA 3175, an NO donor. *Pulmonary Pharmacology and Therapeutics,* *19*(3), 179—188. https://doi.org/10.1016/j.pupt.2005.05.004

Erdoğan, T., Fidan, U., & Özyiğit, G. (2021). Patient-specific tumor and respiratory monitoring phantom design for quality controls of stereotactic ablative body radiotherapy in lung cancer cases. *Physica Medica,* *90*(March), 40—49. https://doi.org/10.1016/j.ejmp.2021.09.003

Flatby, H. M., Rasheed, H., Ravi, A., Thomas, L. F., Liyanarachi, K. V., Afset, J. E., DeWan, A. T., Brumpton, B. M., Hveem, K., Åsvold, B. O., Simonsen, G. S., Furberg, A. S., Damås, J. K., Solligård, E., & Rogne, T. (2021). Risk of lower respiratory tract infections: A genome-wide association study with mendelian randomization analysis in three independent European populations. *Clinical Microbiology and Infection, xxxx.* https://doi.org/10.1016/j.cmi.2021.11.004

Gater, A., Nelsen, L., Coon, C. D., Eremenco, S., O'Quinn, S., Khan, A. H., Eckert, L., Staunton, H., Bonner, N., Hall, R., Krishnan, J. A., Stoloff, S., Schatz, M., Haughney, J., & Coons, S. J. (2021). Asthma daytime symptom diary (ADSD) and asthma nighttime symptom diary (ANSD): Measurement properties of novel patient-reported symptom measures. *Journal of Allergy and Clinical Immunology: In Practice,* 1—11. https://doi.org/10.1016/j.jaip.2021.11.026

Gentilotti, E., De Nardo, P., Cremonini, E., Górska, A., Mazzaferri, F., Canziani, L. M., Hellou, M. M., Olchowski, Y., Poran, I., Leeflang, M., Villacian, J., Goossens, H., Paul, M., & Tacconelli, E. (2022). Diagnostic accuracy of point-of-care tests in acute community-acquired lower respiratory tract infections. A systematic review and meta-analysis. *Clinical Microbiology and Infection,* *28*(1), 13—22. https://doi.org/10.1016/j.cmi.2021.09.025

Ghigna, M. R., & Dorfmüller, P. (2019). Pulmonary vascular disease and pulmonary hypertension. *Diagnostic Histopathology,* *25*(8), 304—312. https://doi.org/10.1016/j.mpdhp.2019.05.002

Gill, S. E., Pape, M. C., & Leco, K. J. (2006). Tissue inhibitor of metalloproteinases 3 regulates extracellular matrix-Cell signaling during bronchiole branching morphogenesis. *Developmental Biology,* *298*(2), 540—554. https://doi.org/10.1016/j.ydbio.2006.07.004

Harrison, O. K., Köchli, L., Marino, S., Luechinger, R., Hennel, F., Brand, K., Hess, A. J., Frässle, S., Iglesias, S., Vinckier, F., Petzschner, F. H., Harrison, S. J., & Stephan, K. E. (2021). Interoception of breathing and its relationship with anxiety. *Neuron,* *109*(24), 4080—4093. https://doi.org/10.1016/j.neuron.2021.09.045. e8.

Hassan Lemjabbar-Alaouia, O. H., Yanga, Y.-W., & Buchanana, P. (2017). Lung cancer: Biology and treatment options. *Physiology & Behavior,* *176*(5), 139—148. https://doi.org/10.1016/j.bbcan.2015.08.002.Lung

Hwang, J. H., Oh, M. R., Hwang, J. H., Choi, E. K., Jung, S. J., Song, E. J., Españo, E., Webby, R. J., Webster, R. G., Kim, J. K., & Chae, S. W. (2021). Effect of processed aloe vera gel on immunogenicity in inactivated quadrivalent influenza vaccine and upper respiratory tract infection in healthy adults: A randomized double-blind placebo-controlled trial. *Phytomedicine, 91.* https://doi.org/10.1016/j.phymed.2021.153668

Iizuka, Y., Nakamura, M., Kozawa, S., Mitsuyoshi, T., Matsuo, Y., & Mizowaki, T. (2019). Tumour volume comparison between 16-row multi-detector computed tomography and 320-row area-detector computed tomography in patients with small lung tumours treated with stereotactic body radiotherapy: Effect of respiratory motion. *European Journal of Radiology,* *117*(May), 120—125. https://doi.org/10.1016/j.ejrad.2019.06.002

Jiménez-Posada, L. D., Maya, J. C., Sánchez-Ocampo, M., López-Isaza, S., Cortes-Ospina, S., Montagut-Ferizzola, Y. J., & Torres, R. (2021). Computational model of trachea-alveoli gas movement during spontaneous breathing. *Respiratory Physiology and Neurobiology,* *294*(May). https://doi.org/10.1016/j.resp.2021.103767

Khaltaev, N., & Axelrod, S. (2019). Chronic respiratory diseases global mortality trends, treatment guidelines, life style modifications, and air pollution: Preliminary analysis. *Journal of Thoracic Disease, 11*(6), 2643–2655. https://doi.org/10.21037/jtd.2019.06.08

Korten, I., Oestreich, M. A., Frey, U., Moeller, A., Jung, A., Spinas, R., Mueller-Suter, D., Trachsel, D., Rochat, I., Spycher, B., Latzin, P., Casaulta, C., & Ramsey, K. (2021). Respiratory symptoms do not reflect functional impairment in early CF lung disease. *Journal of Cystic Fibrosis, 20*(6), 957–964. https://doi.org/10.1016/j.jcf.2021.04.006

Kumar, S., Garg, I. B., & Sethi, G. R. (2019). Serological and molecular detection of Mycoplasma pneumoniae in children with community-acquired lower respiratory tract infections. *Diagnostic Microbiology and Infectious Disease, 95*(1), 5–9. https://doi.org/10.1016/j.diagmicrobio.2019.03.010

Levine, S. M., & Marciniuk, D. D. (2022). Global impact of respiratory disease. In *Chest*. https://doi.org/10.1016/j.chest.2022.01.014

Levine, A. R., Simon, M. A., & Gladwin, M. T. (2019). Pulmonary vascular disease in the setting of heart failure with preserved ejection fraction. *Trends in Cardiovascular Medicine, 29*(4), 207–217. https://doi.org/10.1016/j.tcm.2018.08.005

Lewthwaite, H., Li, P. Z., O'Donnell, D. E., & Jensen, D. (2021). Multidimensional breathlessness response to exercise: Impact of COPD and healthy ageing. *Respiratory Physiology and Neurobiology, 287*(January), 103619. https://doi.org/10.1016/j.resp.2021.103619

Liu, H., Runck, H., Schneider, M., Tong, X., & Stahl, C. A. (2011). Morphometry of subpleural alveoli may be greatly biased by local pressure changes induced by the microscopic device. *Respiratory Physiology and Neurobiology, 178*(2), 283–289. https://doi.org/10.1016/j.resp.2011.06.024

Llucià-Valldeperas, A., de Man, F. S., & Bogaard, H. J. (2021). Adaptation and maladaptation of the right ventricle in pulmonary vascular diseases. *Clinics in Chest Medicine, 42*(1), 179–194. https://doi.org/10.1016/j.ccm.2020.11.010

Lu, X., Gong, J., Dennery, P. A., & Yao, H. (2019). Endothelial-to-mesenchymal transition: Pathogenesis and therapeutic targets for chronic pulmonary and vascular diseases. *Biochemical Pharmacology, 168*(May), 100–107. https://doi.org/10.1016/j.bcp.2019.06.021

Lukarski, D., Stavrov, D., & Stankovski, T. (2022). Variability of cardiorespiratory interactions under different breathing patterns. *Biomedical Signal Processing and Control, 71*(PA), 103152. https://doi.org/10.1016/j.bspc.2021.103152

Maio, S., Baldacci, S., Carrozzi, L., Pistelli, F., Simoni, M., Angino, A., La Grutta, S., Muggeo, V., & Viegi, G. (2019). 18-yr cumulative incidence of respiratory/allergic symptoms/diseases and risk factors in the Pisa epidemiological study. *Respiratory Medicine, 158*(September), 33–41. https://doi.org/10.1016/j.rmed.2019.09.013

Marketos, S. G., & Eftychiades, A. C. (1986). Historical perspectives: Bronchial asthma according to byzantine medicine. *Journal of Asthma, 23*(3), 149–155. https://doi.org/10.3109/02770908609077489

Marshall, B. G., White, V., & Loveridge, J. (2021). Breathlessness and cough in the acute setting. *Medicine (United Kingdom), 49*(2), 93–97. https://doi.org/10.1016/j.mpmed.2020.11.004

Masunaga, A., Takemura, T., Ichiyasu, H., Migiyama, E., Horio, Y., Saeki, S., Hirosako, S., Mori, T., Suzuki, M., Kohrogi, H., & Sakagami, T. (2021). Pathological and clinical relevance of selective recruitment of Langerhans cells in the respiratory bronchioles of smokers. *Respiratory Investigation, 59*(4), 513–521. https://doi.org/10.1016/j.resinv.2021.03.001

Normal versus asthmatic bronchiole: MedlinePlus Medical Encyclopedia Image. (n.d.). https://medlineplus.gov/ency/imagepages/19346.htm

Muiser, S., Gosens, R., van den Berge, M., & Kerstjens, H. A. M. (2022). Understanding the role of long-acting muscarinic antagonists in asthma treatment. *Annals of Allergy, Asthma & Immunology, 128*(4), 352–360. https://doi.org/10.1016/j.anai.2021.12.020

Patria, M. F., & Esposito, S. (2013). Recurrent lower respiratory tract infections in children: A practical approach to diagnosis. *Paediatric Respiratory Reviews, 14*(1), 53–60. https://doi.org/10.1016/j.prrv.2011.11.001

Ramalho, S. H. R., & Shah, A. M. (2021). Lung function and cardiovascular disease: A link. *Trends in Cardiovascular Medicine, 31*(2), 93—98. https://doi.org/10.1016/j.tcm.2019.12.009

Ruscic, K. J., Grabitz, S. D., Rudolph, M. I., & Eikermann, M. (2017). Prevention of respiratory complications of the surgical patient: Actionable plan for continued process improvement. *Current Opinion in Anaesthesiology, 30*(3), 399—408. https://doi.org/10.1097/ACO.0000000000000465

Saers, J., Andersson, L., Janson, C., & Sundh, J. (2021). Respiratory symptoms, lung function, and fraction of exhaled nitric oxide before and after assignment in a desert environment—a cohort study. *Respiratory Medicine, 189*(October), 106643. https://doi.org/10.1016/j.rmed.2021.106643

Sharma, N., Bietar, K., & Stochaj, U. (2022). Targeting nanoparticles to malignant tumors. *Biochimica et Biophysica Acta (BBA) - Reviews on Cancer, 1877*(3), 188703. https://doi.org/10.1016/j.bbcan.2022.188703

Shi, T., Pan, J., Vasileiou, E., Robertson, C., & Sheikh, A. (2022). Risk of serious COVID-19 outcomes among adults with asthma in scotland: A national incident cohort study. *The Lancet Respiratory Medicine, 2600*(21), 1—8. https://doi.org/10.1016/s2213-2600(21)00543-9

Soriano, J. B., Kendrick, P. J., Paulson, K. R., Gupta, V., Abrams, E. M., Adedoyin, R. A., Adhikari, T. B., Advani, S. M., Agrawal, A., Ahmadian, E., Alahdab, F., Aljunid, S. M., Altirkawi, K. A., Alvis-Guzman, N., Anber, N. H., Andrei, C. L., Anjomshoa, M., Ansari, F., Antó, J. M., ... Vos, T. (2020). Prevalence and attributable health burden of chronic respiratory diseases, 1990—2017: A systematic analysis for the global burden of disease study 2017. *The Lancet Respiratory Medicine, 8*(6), 585—596. https://doi.org/10.1016/S2213-2600(20)30105-3

Sorino, C., Mondoni, M., Lococo, F., Marchetti, G., & Feller-Kopman, D. (2022). Optimizing the management of complicated pleural effusion: From intrapleural agents to surgery. *Respiratory Medicine, 191*(November 2021), 106706. https://doi.org/10.1016/j.rmed.2021.106706

Staso, P., Wu, P., Laidlaw, T., & Cahill, K. (2021). Assessing systemic symptoms during aspirin challenge in aspirin exacerbated respiratory disease (AERD). *Journal of Allergy and Clinical Immunology, 147*(2), AB39. https://doi.org/10.1016/j.jaci.2020.12.174

Suarez, C. J., Dintzis, S. M., & Frevert, C. W. (2012). Respiratory. In *Comparative anatomy and histology* (pp. 121—134). https://doi.org/10.1016/B978-0-12-381361-9.00009-3

Sung, J. H., Brown, M. C., Perez-Cosio, A., Pratt, L., Houad, J., Liang, M., Gill, G., Moradian, S., Liu, G., & Howell, D. (2020). Acceptability and accuracy of patient-reported outcome measures (PROMs) for surveillance of breathlessness in routine lung cancer care: A mixed-method study. *Lung Cancer, 147*(June), 1—11. https://doi.org/10.1016/j.lungcan.2020.06.028

Syed, M. N., Shah, M., Shin, D. B., Wan, M. T., Winthrop, K. L., & Gelfand, J. M. (2021). Effect of anti—tumor necrosis factor therapy on the risk of respiratory tract infections and related symptoms in patients with psoriasis—a meta-estimate of pivotal phase 3 trials relevant to decision making during the COVID-19 pandemic. *Journal of the American Academy of Dermatology, 84*(1), 161—163. https://doi.org/10.1016/j.jaad.2020.08.095

Vardakas, K. Z., Mavroudis, A. D., Georgiou, M., & Falagas, M. E. (2018). Intravenous plus inhaled versus intravenous colistin monotherapy for lower respiratory tract infections: A systematic review and meta-analysis. *Journal of Infection, 76*(4), 321—327. https://doi.org/10.1016/j.jinf.2018.02.002

Waghmare, A., Xie, H., Kuypers, J., Sorror, M. L., Jerome, K. R., Englund, J. A., Boeckh, M., & Leisenring, W. M. (2019). Human rhinovirus infections in hematopoietic cell transplant recipients: Risk score for progression to lower respiratory tract infection. *Biology of Blood and Marrow Transplantation, 25*(5), 1011—1021. https://doi.org/10.1016/j.bbmt.2018.12.005

Wang, L., Fang, R., Zhu, M., Qin, N., Wang, Y., Fan, J., Sun, Q., Ji, M., Fan, X., Xie, J., Ma, H., & Dai, J. (2021). Integrated gene-based and pathway analyses using UK Biobank data identify novel genes for chronic respiratory diseases. *Gene, 767,* 145287. https://doi.org/10.1016/j.gene.2020.145287

Witt, W. P., Weiss, A. J., & Elixhauser, A. (2014). Overview of hospital stays for children in the United States, 2012. *HCUP Statistical Brief #187, 167*(2), 1—17.

Won, C. H. J., & Kryger, M. (2014). Sleep in patients with restrictive lung disease. *Clinics in Chest Medicine, 35*(3), 505—512. https://doi.org/10.1016/j.ccm.2014.06.006

World Health Organization. (2007). *Global surveillance, prevention and control of chronic respiratory diseases: A comprehensive approach.*

Wu, E. K., Eliseeva, S., Rahimi, H., Schwarz, E. M., & Georas, S. N. (2019). Restrictive lung disease in TNF-transgenic mice: Correlation of pulmonary function testing and micro-CT imaging. *Experimental Lung Research, 45*(7), 175—187. https://doi.org/10.1080/01902148.2019.1636899

Xu, R., Liu, P., Zhang, T., Wu, Q., Zeng, M., Ma, Y., Jin, X., Xu, J., Zhang, Z., & Zhang, C. (2021). Progressive deterioration of the upper respiratory tract and the gut microbiomes in children during the early infection stages of COVID-19. *Journal of Genetics and Genomics, 48*(9), 803—814. https://doi.org/10.1016/j.jgg.2021.05.004

Yang, F., Busby, J., Heaney, L. G., Menzies-Gow, A., Pfeffer, P. E., Jackson, D. J., Mansur, A. H., Siddiqui, S., Brightling, C. E., Niven, R., Thomson, N. C., Chaudhuri, R., Patel, M., Gore, R., Brown, T., Vyas, A., & Subramanian, D. (2021). Factors associated with frequent exacerbations in the UK severe asthma registry. *Journal of Allergy and Clinical Immunology: In Practice, 9*(7), 2691—2701. https://doi.org/10.1016/j.jaip.2020.12.062. e1.

Yang, H., Gao, R., Li, J., Lian, W., Guo, B., Wang, W., & Huang, H. (2022). Dynamic response mechanism of borehole breathing in fractured formations. *Energy Reports, 8,* 3360—3374. https://doi.org/10.1016/j.egyr.2022.02.148

Zengel, P. (2019). Sinusitis, Otitis, Laryngitis und Co.: Wann Antibiotika verschreiben? *MMW - Fortschritte Der Medizin, 161*(1), 40—44. https://doi.org/10.1007/s15006-019-0066-y

Ziou, M., Tham, R., Wheeler, A. J., Zosky, G. R., Stephens, N., & Johnston, F. H. (2022). Outdoor particulate matter exposure and upper respiratory tract infections in children and adolescents: A systematic review and meta-analysis. *Environmental Research, 210*(February), 112969. https://doi.org/10.1016/j.envres.2022.112969

CHAPTER TWO

Asthma: The disease and issues in monitoring the asthmatic attack

Asthma is a heterogeneous, multifactorial disease with variable and mostly reversible obstruction of the respiratory passages based on a chronic bronchial inflammatory reaction. The disease itself is highly variable with episodes of attacks of differing severity interspersed between symptom-free periods. The symptoms (cough, rhonchi, wheezing, chest tightness, or shortness of breath) are vary between patients but generally correlated with expiratory flow limitation, which occurs in episodes. Often, between episodes, asthmatic patients are totally symptom-free and appear in full health. However, for a significant proportion of asthmatics, the disease lingers on and if poorly controlled, chronic asthma disease, irreversible airway obstruction, and persistent symptoms become common. This chapter elaborates asthma as one of the severe respiratory illness and its symptoms, pathology, and treatment with capnography technology.

2.1 Asthma: the burden of disease

Asthma was first described by the ancient Greek physician Hippocrates and derived from the Greek word asthma meaning panting or gasping. It is a heterogeneous disease with complex pathophysiology and phenotype. The word "asthma" originates from the Greek meaning short of breath, meaning that any patient with breathlessness was asthmatic. Since ancient times, considerable advances have been made in understanding the genetics, epidemiology, and pathophysiology of asthma, a condition that has increased in prevalence worldwide over the past 20 years (Ortiz & Sanders, 2012). Asthma remains one of the most common chronic diseases in the world. It affects people from all parts of the world, regardless of geography, climate, environment, socioeconomic status, race, or religion. It is a common chronic disorder of the airways that is complex and characterized by variable and recurring symptoms, airflow obstruction, bronchial hyperresponsiveness, and an underlying inflammation. These episodes are usually associated with widespread, but variable, airflow obstruction within lung that is often reversible either spontaneously or with the treatment. Its impact on all our

Systems and Signal Processing of Capnography as a Diagnostic Tool for Asthma Assessment
ISBN: 978-0-323-85747-5
https://doi.org/10.1016/B978-0-323-85747-5.00007-3

communities ranges from its adverse impact on quality of life, to its significant healthcare cost up to the continuing mortality due to the disease every year (Cukic et al., 2012). Each day, hundreds around the world still die from this disease; unfortunately, the poor developing world still bears the brunt of the disease, reporting 80% of the world's asthma-related deaths (Holgate & Thomas, 2017).

Asthma is a serious global health problem, and there are approximately 339 million people worldwide who have asthma with over 80% of asthma-related deaths occur in low- and lower-middle-income countries (Bouazza et al., 2021). Globally, asthma is ranked 16th among the leading causes of years lived with disability and 28th among the leading causes of burden of disease, as measured by disability-adjusted life years, and it is likely that by 2025, a further 100 million may be affected. The incidence among adults varies widely throughout the world, reportedly as high as 12 cases per 1000 person/years. In the 1950 and 1960s, there was a marked increase in the number of asthma patients in the Western countries as industrialization and its accompanying air pollution caused a worrying impact (WHO, 2021). Since then, industrialization has occurred throughout the world, and asthma rates have spiked in all these areas from the resulting air pollution. As we understand more about the disease itself, and how air pollution plays a significant part of the disease, many countries that have worked to control their air pollution problems have now reported reducing numbers of patients. Other countries however have not, and children and the poor of those countries continue to suffer the effects of this.

There is a large geographical variation in asthma prevalence, severity, and mortality. While asthma prevalence is higher in high-income countries, most asthma-related mortality occurs in low-middle-income countries (Dharmage et al., 2019). During the second half of the 20th century, notably since the 1960s, a sharp increase in asthma prevalence was observed in several developed countries. This observation was a result of repeated cross-sectional surveys of prevalence of asthma, mainly in children but also in adults. As a result of this observation, in the 1990s, a series of epidemiological studies were established across the world to estimate global asthma prevalence and incidence and identify risk factors associated with these outcomes. These include large multinational studies in children and in adults. These studies confirmed that asthma is one of the most common chronic diseases across the globe in all age groups, and there is substantial variation in asthma prevalence worldwide.

Table 2.1 Risk factors for asthma.

Risk factors for asthma	Host factors
	Genetic predisposition
	Atopy (allergy)
	Infections (usually viral)
	Gender/Race (ethnicity)
	Stress-related
	Obesity
	Environmental factors
	Indoor contaminants (dust, kitchen smoke, pets)
	Outdoor pollution
	Tobacco smoke
	Socioeconomic factors
	Occupational exposures
	Environmental changes, weather variance

Most of the asthma starts in early childhood. About 95% of asthma patients have their first episode before the age of 6 years. Young males are consistently reported to have more prevalent asthma than young females. However, by the ages of 11—16 years, the prevalence of asthma in young females starts to exceed those of young males. The main risk factors associated with childhood-onset asthma are genetic predisposition, a family history of allergy and asthma, recurrent viral respiratory infections, allergic sensitization, and tobacco smoke exposure, as shown in Table 2.1 (de Nijs et al., 2013). Based on a statistical study conducted by Global Initiative for Asthma, the global prevalence of asthma in children is above 50% from the sum of asthma cases reported in 2017 (Fig. 2.1). A significant proportion of children with childhood asthma do not go on to suffer from any asthmatic attacks in their adulthood. But for others, asthma remains part of their life as they grow into adults.

Genetics are known to play a role, with asthma with heritability ranging between 35% and 95%. Large genetic studies have identified hundreds of genetic variants associated with an increased risk of asthma. Epigenetic variations in how the genetic code is translated have also been shown to have a role in the development of asthma. The identified genes responsible are more than 100, and many polymorphisms have been shown to be associated to the onset of asthma, although none of these, alone or in combination, is able to predict the occurrence of disease (Miraglia del Giudice et al., 2014). The environmental factors most involved in the onset of asthma in children

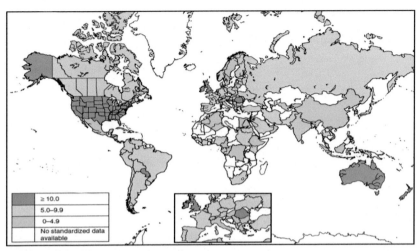

Figure 2.1 Global prevalence of asthma. *Reproduced with permission from (Holgate & Thomas, 2017) Copyright 2017 Elsevier B.V.*

are represented by allergens, tobacco smoke, respiratory infections, and air pollution. Indoor allergens (dust mites, mold, and animal dander) and outdoor (pollens and molds) are able to induce sensitization by prolonged exposure and trigger acute asthma. Allergic sensitization, in the concept of atopic march, represents a major risk factor for the development of asthma. In particular, the subjects polysensitized and with food allergy may present more severe asthma. The exposure to cigarette smoke in both prenatal and postnatal increases the risk of the child becoming asthmatic and the asthma severity. It has also noted recently that obesity is a risk factor for asthma because it causes an increase of leptin, TNF-α, and IL-6, which exert a proinflammatory noneosinophil action. In addition, the lack of physical activity, for weight gain, contributes to the determinism of the disease. Respiratory infections, especially viral infections early in life, increase the risk of developing asthma, particularly if the symptoms are severe (MD, 2015).

Despite significant improvements in the diagnosis and management of asthma over the past decade, as well as the availability of comprehensive and widely accepted national and international clinical practice guidelines for the disease, asthma control remains suboptimal (Quirt et al., 2018). It is appearing to be the most common chronic disease of childhood, but people of all ages are affected by this illness that, when uncontrolled, can place severe limits on daily life and is sometimes fatal. A thorough history and physical examination along with are important for the diagnosis of asthma.

The prevalence of asthma is high in most countries as it appears to be a significant cause of morbidity in patients of all ages (Cukic et al., 2012; Dharmage et al., 2019). This study therefore aims to identify and review current studies on the pathophysiology, diagnosis, and appropriate treatment of asthma.

2.2 The pathophysiology of asthma attack

At its core, asthma is an inflammatory disease characterized by inflammatory exudates, bronchoconstriction, and mucous plugging causing narrowing of the breathing passages. The affected breathing passages are typically the smaller bronchioles near the level of the alveoli, which are typically very small and may easily be blocked significantly or totally. This narrowing usually occurs acutely over a short period of time, an exacerbation of asthma or commonly referred to as the "asthmatic attack." Currently, severe narrowing of the bronchioles occurs as a result of several concurrent events. Firstly, the small bronchioles constrict leading to narrowing. Then, the mucous membrane swells due to edema, a collecting of fluid within the cells, thus narrowing the bronchioles further. Lastly, an excess of secretions and exudates from the underlying inflammatory process then often block up the narrowed bronchiole totally. The pressures within the chest cavity, which swings from negative to positive during the inspiratory–expiratory breath cycle, means that the bronchioles are naturally most narrowed during exhalation. The overall consequence of the entire process is the obstruction of airflow through the narrowed small bronchioles, particularly during exhalation.

We recall from the earlier discussion that exhalation is normally a passive process, resulting from a relaxation of the muscles of the diaphragm and chest wall, together with the elastic recoil of the lungs. During asthmatic attacks, exhalation is particularly difficult due to the difficulty of moving air through the narrowed bronchioles. Thereby, asthmatic patients are often forced to use active muscular activity during exhalation to force air through those narrowed air passages. The forceful movement of air through the narrowed air passages causes a whistle-like situation resulting in the typical wheeze of asthmatic attacks. The schematic illustration in Fig. 2.2 represents the asthma pathogenesis pathway and its effect in changing the bronchioles during exacerbation of asthma.

In mild attacks, the body compensates by increasing its muscular activity both during inhalation and exhalation to aid breathing efforts. The work of

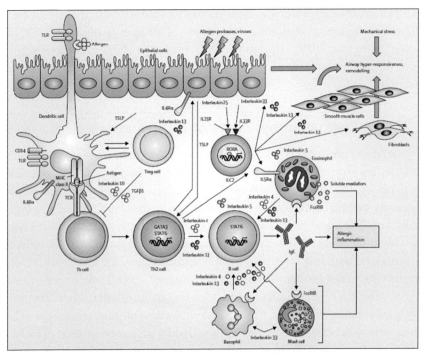

Figure 2.2 Changes in the bronchioles during exacerbation of asthma. *Reproduced with permission from (Martinez & Vercelli, 2013) Copyright 2013 Elsevier B.V.*

breathing-in faster is in hope of increasing oxygen brought into the body to meet the higher demand. The added work of breathing out aims to clear the air and CO_2 from the alveoli, so that the next breath can fill the alveoli with refreshed air. These increases in rate and depth of breathing compensates for the initial impacts on O_2— CO_2 exchange due to the narrowed bronchioles. Patients may feel slightly breathless, or feel tightness in the chest, cough and have a slight wheeze but may often continue with some daily activities.

In severe asthmatic attacks, however, more and more air passages become narrowed up to the point of getting obstructed totally, and areas of the lungs become totally cut off by the blocked bronchioles, affecting the overall ability of the lungs to exchange O_2 and CO_2. In some parts of the lungs, no new air comes into the alveoli and O_2–CO_2 exchange grinds to a halt. This leads to marked ventilation: perfusion mismatches; a hallmark of severe asthmatic attacks. Patients becoming more and more breathless as they struggle to breath to bring in as much O_2 and throw out as much CO_2 as they can. Extensive muscle activity is now needed just to inhale and exhale. All that muscular activity has a severe metabolic cost, i.e., the muscles themselves require oxygen to continue to function, thereby causing a further increase in the need for oxygen at the time when its supply is already threatened. This becomes a vicious cycle, and if it is not rapidly corrected or reversed, it will lead at some point to a loss of those muscular functions, which is followed by a rapid deterioration of respiratory efforts. Blood oxygen levels precipitously decline causing vital organ dysfunctions, loss of consciousness, and potentially a rapid loss of life. Early diagnosis of asthmatic attack is therefore important. Initiating treatment early breaks the development of the vicious cycle and allows the body to recover quickly. Early assessment of the severity of the attack is also important to guide treatment. Therefore, careful and repeated assessment of patients with an asthmatic attack is key to their proper management, as their status may change dramatically within hours. Such is the precocious nature of the asthmatic attack.

2.3 Types of asthma

2.3.1 Allergic asthma

Inhaling allergens is the most common trigger for inducing allergic asthma. Allergens may include dust mites, pet dander, pollen, or mold (Prevention, 2020). Sometimes, allergic asthma may be due to patients' occupations (paint, metal, or gas); less commonly food allergens are responsible of exacerbations of asthma. Therefore, patients may want to know the primary cause and the exact trigger in order to avoid or reduce their exposure. It is often necessary to consult an allergist or immunologist for skin testing to identify potential allergens.

2.3.2 Reactive asthma

While asthma patients are able to identify their causative allergen sometimes, e.g., some patients may experience asthma attacks during spring when the pollen counts are high, more frequently they are not able to identify a specific allergen. They may experience attacks in response to changes in temperature, e.g., patients who experience attacks mostly only during the winter or cold season (Fernández-Nieto et al., 2006) or when the environmental pollution index is high.

2.3.3 Exercise-induced asthma

Scoggins described exercise-induced asthma as a condition, whereby symptoms of asthma could be exacerbated after performing exercise. Exercise-induced asthma was characterized by fall in forced expiratory volume in the first second (FEV1) or peak expiratory rate (PEF) greater than 10% of preexercise values. The research further elaborated that the second phase of exercise-induced asthma was severer, which begins after the exercise and lasts about half an hour to an hour. Exercise-induced asthma can occur in all patients who have been diagnosed with types of asthma. In a study conducted in a laboratory, 70%–80% of people who have asthma provoked by exercise showed a significant reduction of airflow, typical to exercise-induced (Côté et al., 2018). The following two hypotheses were proposed as the mechanism in exercise induced asthma, the osmotic hypothesis, and the thermal hypothesis. The former suggested that changes in water and temperature in the airways during exercise will cause bronchoconstriction. The airways underwent a conditioning process (warming and humidifying the air) as large air volumes move in and out of the lungs. When performing exercise, there was increase in the rate of ventilation, by a factor of 20 (Rodriguez Bauza & Silveyra, 2021). There is conditioning of air, which traveled from the upper airways to the lower airways. More release of heat and water from the airway cells is required to heat and humidity the air. When exercise stops, the ventilation decreased, subsequently the airways to rewarm rapidly as there are no more loss of heat and water to the air. Bronchoconstriction of the airways resulted from this cycle of cooling and rewarming of the airway. A study to support this theory was conduct where breathing warm and humidified air ameliorated exercise-induced broncho-constriction and breathing of cold air worsened the situation (Boonpiyathad et al., 2019). The latter, osmotic hypothesis explains that airway dehydration during exercise contributes to events leading to airway smooth muscle contractions and increase in airway resistance. There was large movement of air in and out of the lungs during exercise, along with increased respiratory rate and tidal volume (Lang, 2019). Loss of water from the airways was increased through evaporation. Osmolarity was increased in the cells causing release of inflammatory chemicals, which lead to contraction of smooth muscle cells.

2.3.4 Nonallergic asthma

According to the Robinson's Basic Pathology, nonallergic asthma or nonatopic asthma or intrinsic asthma is a form of asthma that has no evidence of allergen sensitization (Nema, 2004). 10%—33% of asthmatic patients are of this type. The main characteristics of nonallergic asthma are negative skin prick or in vitro—specific IgE tests to a panel of local allergens, and at a minimum, a panel of perennial allergens; total serum IgE levels are typically low (<150 IU/mL) (Pakkasela et al., 2020). A positive family history is less important in this type of asthma. The main triggers are viral respiratory infections and inhaled air pollutant. The viral infection to respiratory mucosa lowers the threshold of the subepithelial vagal receptors to irritants. There are two major mechanisms that postulate neutrophilic inflammation, which are dysregulated neutrophil-mediated immune responses due to infections or defects in the resolution of inflammations and the activation of IL-17 pathway (Peters, 2014). The mechanism of dysregulated neutrophil-mediated immune responses does not respond from the normal TH2 inflammatory mechanism as seen in allergic asthma. Endotoxins, soluble fragments of lipopolysaccharides from the outer membrane of Gram-negative bacteria and Gram-positive cell wall components such as lipoteichoic acids act as pathogen-associated molecular patterns (PAMPs) that are recognized by Toll-like receptors, CD-14, and collectins. This will create a cascade of release of inflammatory mediators and therefore present in sputum, bronchoalveolar fluid, and plasma (Fang et al., 2020). This led to the normal obstructive signs and symptoms of asthma. Another theory that suggested nonallergic asthma because of defects in the resolution of inflammations was supported by a study by Uddin et al. (2010), which proved that in a severer form of asthma in patients who have neutrophilia, i.e., showed high significant count ($P = .008$) toward decreasing number of apoptotic neutrophils (Baos et al., 2018).

2.3.5 Occupational asthma

Occupational asthma has been reported in 10% of workers exposed to sensitizing agents in cross-sectional studies. According to Tarlo & Lemiere (2014), occupational asthma can be divided into two main groups, which are sensitizer-induced asthma and nonsensitizing, irritant-induced occupational asthma (Friedlander et al., 2014). Sensitizer-induced asthma is caused by specific workplace sensitizers that induce asthma mechanism associated with specific molecular response. The sensitizers are commonly high-molecular-weight agents (>10kD, usually protein and glycopeptide). They can initiate production of specific IgE antibodies and thus typical immune response. Low-molecular-weight chemicals can also induce sensitization and thus asthma. Some examples of chemicals are complex platinum salts (in platinum refineries and in manufacturing of cytotoxic drugs), rhodium salts (used in electroplating), salts of nickel, chrome and cobalt, and acid anhydrides (used as hardeners in epoxy resins in chemical plants and in powder paints). While diisocyanates, which are used in flexible polyurethane foam, are the most common cause of asthma in industrialized area (Jarvis et al., 2005). Nonsensitizing-induced occupational asthma is caused by exposure of agents considered to be airway irritants in the absence of sensitization. It could be caused by alkaline dust as shown from the World Trade Center disaster in 2001. A study showed 16% of persons who were highly exposed at a proximity to the disaster to have irritant-induced asthma (Cartier, 2015). Some reports consider the likelihood of low-dose airborne irritating chemicals can cause occupational asthma. Previous atopic disease and long-term exposure of irritants may exacerbate underlying hyperresponsiveness or coincidental onset of asthma. Workers who are constantly exposed to cleaners (either industrial or domestic) are likely to get nonsensitizing-induced occupational asthma. Those include cleaners, nurses, textile workers, hog farmers, poultry workers and aluminum portroom workers are highly susceptible to this disease (Quirce & Sastre, 2019; Tiotiu et al., 2020). The pathophysiological mechanism of occupational asthma is similar non-work-related asthma, including IgE-dependent mechanism associated with high-molecular-weight sensitizers and some low-molecular-weight sensitizers. However, some studies have shown that for asthma that is induced by low-molecular-weight substance such as diisocyanates and for irritant-induced asthma, the mechanisms are completely delineated (Choi et al., 2019; Jarvis et al., 2005).

2.4 Classification in chronic asthma

According to Kwah and Peters (2019), the classification of asthma severity in patients who are more than 12 years old is divided into intermittent, mild persistent, moderate persistent, and severe persistent. This classification is based on the signs and symptoms of the patient based on the duration and the timing when they have the symptoms (Kwah & Peters, 2019). The importance of this classification is to determine the method of control of asthma in categorized patients. Persistent asthma is an indicator for preventive treatment. Meanwhile, current protocol from GINA states that asthma severity is assessed retrospectively from the level of treatment required to control symptoms and exacerbations (Brown et al., 2002). A questionnaire is given to patients to assess symptoms for the past 4 weeks.

Questions asked are frequency daytime symptoms, presence of nocturnal symptom, frequency of the patient taking short-acting beta-agonist as reliever to symptoms, and any limitations of activity due to asthma. These questions will stratify the patient into levels of control, which are whether they are well-controlled, partly controlled, or uncontrolled (Maike Grotheer et al., 2019).

2.4.1 Intermittent

Patients with intermittent asthma may have either daytime symptoms (which are cough, wheeze, dyspnea, and heavy breathing), and/or nocturnal symptoms (which includes cough, wheeze, dyspnea, awakenings, and tiredness during the day) (Fergeson et al., 2017). For the diagnosis of the patient having intermittent asthma, daytime symptoms must occur not more than twice per week, and nocturnal symptoms occur less or equal to twice per month. In between the attacks, the patient is normal without any symptoms, without any restrictions in their normal activity.

2.4.2 Persistent asthma

There are three grades of persistent asthma. In mild persistent asthma, flare-ups may occur more than twice a week during the day, but less than once per day. Nighttime flare-ups may occur more than twice a month, but should be less than once a week. Attacks may affect normal activity. Any measure of lung function will be more than 80% (Halpin, 2020; Kuprys-Lipinska et al., 2020). For moderate persistent asthma, flare-ups occur daily and may last several days. Nighttime flare-ups occur more than once a week often disturbing sleep. Measures of lung function may record between 60% and 80% of normal levels. Patients with severe persistent asthma have daily and frequent symptoms that affect sleep and normal daily activities. Lung function tests are often less than 60% of normal levels, and there is extremely limitation in normal activity (Page, 2012).

2.5 Symptoms and signs of asthma

The four main symptoms of asthma are wheezing, coughing, shortness of breath, and subjective sensation of tightness in the chest. According to the Malaysian Pediatrics Protocol, features suggestive of asthma in children younger than 5 years old are cough, typically dry cough, wheezing, difficult or heavy breathing or shortness of breath, reduced activity, having a past history or family history of allergic disease or asthma in first degree relatives (Ismail et al., 2019). From a clinician perspective, descriptive symptoms help in diagnosing someone with asthma. History taking should be taken in detail and personal, family, and history should be sought thoroughly so as not misdiagnose asthma with different diseases (Patadia et al., 2014). For each symptom, the intensity, duration, frequency, environmental exposure, nocturnal frequency, and seasonal component should be assessed.

2.5.1 Wheezing

Expiratory wheeze is the hallmark of acute asthma exacerbation (Kwong & Bacharier, 2019). Wheezing is defined as a musical, high-pitched, whistling sound produced by airflow turbulence, due to airflow passing through narrow bronchioles. Wheezing in pediatric patients is usually recurrent and occurs during sleep or with triggers such as physical activity, laughing, crying, or exposure to tobacco smoke or air pollution (Fainardi et al., 2020). Wheezing can determine the extent of severity in asthma in acute exacerbations. In mild asthma, only end-expiratory wheezing is present. Meanwhile, in severe asthma, both inspiratory and expiratory wheezing is present. In the severest form (life-threatening asthma), absence, or loss in wheezing will occur due to limited airflow or significantly narrowed airway (Le Souëf, 2018). This denotes impending respiratory failure and respiratory muscle fatigue. Diagnosing patient with wheezing to be asthmatic should be done after analyzing symptoms, physical examination, and diagnostic testing (which are spirometry and bronchodilator trial), By this way, other differential diagnoses for local bronchial narrowing can be ruled out such as foreign body or structural obstruction, especially if the wheezing is monophasic and begins and ends at the same point in each respiratory cycle (Bacharier et al., 2021).

2.5.2 Cough and mucous production

Cough is another common symptom for asthma. Sometimes, cough is the only symptom present in an asthmatic patient and commonly occurs at night and usually nonproductive. Sometimes, sputum is present, and when it does, it will appear clear mucoid or pale yellow (Ogawa et al., 2014). If cough is the only symptom, typically in adult patients, cough could be analyzed based on either it occurs seasonally or after exposure to triggers, to diagnose it as asthma. Histologically, the number of neutrophils in sputum can indicate infectious disease as the trigger of coughing in asthma (McCallion & De Soyza,

2017). Cough in pediatric patients is typically recurrent or persistent, nonproductive cough and worsens at night. It may be accompanied by wheezing or breathlessness. In the absence of respiratory infections, usually cough will be triggered with laughing, crying, or exposure to tobacco smoke (Vigeland et al., 2017).

2.5.3 Chest tightness

Chest tightness is a subjective symptom in asthma, in a way that it differs from an individual view to another. It is common mainly in exercise-induced asthma and nocturnal asthma (Edmondstone, 1998). Generalized chest pain, sensation of chest congestion (typically feeling a band-like constriction), or difficulty in inspiration are commonly referred by patients who suffer from chest tightness. Chest tightness is nonspecific, patients having this symptom should be differentiated from other diseases of different systems of the body (Arora & Bittner, 2015).

2.5.4 Dyspnea

Dyspnea is also a subjective sensation of shortness of breath. Dyspnea is non-diagnostic for asthma and further work-up should be done as means for diagnosis (Campbell, 2017). However, the characteristics that patients with asthma usually describe are increase in "work" or "tightness in the chest" (O'Donnell et al., 2020).

2.6 Assessment of asthma

Assessment of the severity of the asthmatic attack relies on the symptoms experienced by the patient, presence of certain physical signs, and the measurement of vital parameters. Common symptoms of the asthmatic attack are cough, wheezing, chest tightness, and difficulty breathing especially during exhalation. Cough is often an early symptom and may be unrecognized as an early sign of an impending asthmatic attack especially in young children. For example, young children who cough at night when the weather turns colder may have asthma (AsthmaUK, 2020). Wheezing is the most recognizable marker of asthma. In mild exacerbations of asthma, the wheeze may be heard predominantly only during the expiratory, or end-expiratory phase. In more severe disease, it may be heard during both inspiratory and expiratory phases. But sometimes in the most severe asthmatic attacks, the wheeze disappears, as the bronchioles become so severely blocked that no air moves through them anymore, and no further sounds are produced. So, the absence of a wheeze in itself may not be a good sign. Chest tightness and the complaints of difficult breathing are very common symptoms experienced by patients during the asthmatic attack. However, each patient perceives these symptoms differently. Many studies have shown that the presence of these symptoms does not reliably indicate the severity of the attack (Atta et al., 2004). Many patients complain of relatively mild symptoms despite having severe obstruction of the breathing passages; on the other hand, other patients may complain of marked symptoms of chest tightness and difficulty in breathing although their narrowing of bronchioles is minimal (Bijl-Hofland et al., 1999). The history provided by patients, of the symptoms they have, correlates poorly with the severity of the asthmatic attack; and should not be relied upon solely to assess severity of the attack.

Physical examination findings (e.g., heart rate, respiratory rates, pulse oximetry readings, level of consciousness, use of accessory muscles of respiration) provide better estimation of the severity of the attack. In mild to moderate attacks, the compensatory breathing efforts by the patient allows for normal O_2-CO_2 exchange without too much added effort. This can be seen as a relatively normal physical examination findings and vital sign measurements. As the disease worsens, increases in heart rate and respiratory rates become more prominent. In severe attacks though, physical examination will reveal the significant muscular effort undertaken to compensate for the respiratory distress, as well as the underlying abnormalities in the O_2-CO_2 exchange. The most severe life-threatening conditions usually show signs of vital organ failure, markedly abnormal vital signs suggesting impending collapse. Fig. 2.3 shows the standard protocol applied in identification and treatment of asthma.

Since asthma is primarily about airflow obstruction through narrowed bronchioles, the best test of asthma severity would be a direct measurement of the degree of narrowing of the bronchioles. Unfortunately, this single-best test does not yet exist. There is no way of directly measuring diameter of the small bronchioles during the asthmatic attack. However, there are spirometry tests to measure the degree of overall airflow that the patient can generate. The primary test for asthma is known as the peak expiratory flowrate (PEF). This test measures a forceful maximal expiration performed by the patient, and as measured as the maximal flow rate that can be generated during expiration in L/min. A normal PEF shows that the bronchioles are normally open and allowing full airflow through them. On the other

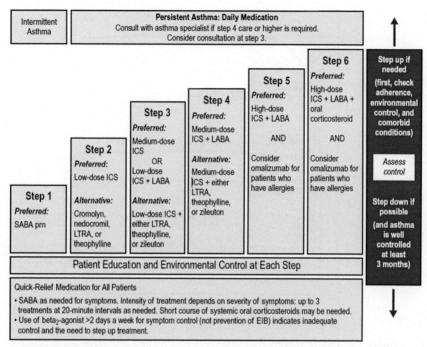

Figure 2.3 Example of an asthma severity score and treatment protocol. *Reproduced with permission from (Holgate & Thomas, 2017) Copyright 2017 Elsevier B.V.*

hand, in asthma, the constricted bronchioles will be narrowed, and the patient will be unable to generate a normal PEF flow. The lower the PEF reading, the more severe the disease. The PEF reading therefore should give clinicians a better idea of severity of the asthmatic attack. Similarly, the first forced expiratory volume (FEV1) test, which measures the total expired volume in 1 second, also indicates the degree of narrowing of the bronchioles, and the severity of the attack. The FEV1 changes in asthma are shown below. Fig. 2.4 illustrates the assessment of asthma in asthmatic and healthy patients, based on the forced vital capacity (FVC) and FEV1 trends. In summary, the assessment of the severity of the asthmatic attack is an important step in the appropriate management of asthma. There is no single-best marker for severity assessment. But a combination of symptoms, physical findings, and measured parameters will give a much better assessment of severity. This has led to the use of clinical scoring systems (e.g., asthma severity scores) that try to correlate symptoms and clinical findings with severity, which in turn then helps to guide treatment.

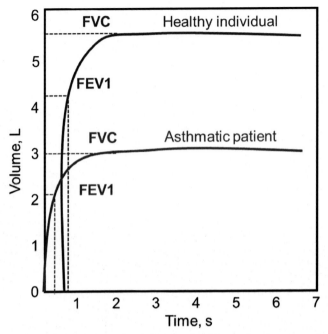

Figure 2.4 Typical spirometry readings in healthy and asthmatic patients.

2.7 Treatment modalities for the asthmatic attack

The main treatment modality of the acute asthmatic attack is short-acting bronchodilators (SABA), steroids, and oxygen support (Saadeh, 2020). SABAs are drugs that reverse the bronchoconstriction and relieve the narrowing of the breathing passages. SABAs are the primary reliever medication provided for the acute asthmatic attack. These drugs are generally most effective when inhaled directly into the lungs, using nebulization or in inhaler devices. This allows the drugs to act directly on the areas of bronchoconstriction, reducing the degree of narrowing and relieving symptoms. Apart from the reliever SABA medications, steroids are primary controller medication used in acute asthmatic attacks. Steroids are used to reduce the inflammatory process, reduce the severity of the attack, and reduce the frequency and duration of the attack. Steroids do not act immediately and therefore play no immediate part in relieving the symptoms of patients with an asthmatic attack; but they are vital in reducing the overall length and severity of the attack on the whole. Oxygen supplementation is commonly used to improve the delivery of oxygen to the alveoli, improving blood oxygenation levels.

For more severe patients, continuous or more intensive nebulization of medications is often used. However, as many of the bronchioles may have become totally blocked, the inhaled or nebulized medications may not be able to reach the target bronchioles, in order to affect its intended outcome. So, other routes of medications are used instead. Intravenous bronchodilators are used among them SABAs, anticholinergics, and smooth muscle relaxants. In life-threatening asthma, the most important step is often to take over the ventilation from the tiring patient, ensuring adequate oxygenation using a ventilator. Occasionally, Heliox, an oxygen—helium mixture, is used to aid delivery of oxygen through highly constricted bronchioles.

2.8 Problems with the monitoring of asthmatic patients

It should be clear by now that asthma can be deadly. It may start suddenly and acutely, and patients may precipitously decline into life-threatening conditions. On the other hand, with medications, some patients improve rapidly; others may need more time and repeated doses. It is therefore essential to be able to monitor, not just the severity of the asthmatic attack, but the continuing status of the patient so that their treatment can be tailored according to their needs at that time. While the asthma severity scores are useful as an initial tool to determine asthma severity, they are performed very poorly as monitoring tools for asthma. There are several reasons for this. SABAs cause increased heart rate, increased respiratory rates, anxiety, palpitations, and tremors. So, patients who have been provided with SABAs often have these same physical findings, which according to the asthma severity score suggest deterioration, but are in fact because of the administered drugs. Oxygen supplementation is often provided to asthmatic patients to help reduce their discomfort, but this renders monitoring of oxygen levels by pulse oximetry less sensitive to identifying deterioration. Ultimately, the asthma severity scores, which have been the mainstay of assessment of severity of diseases, are only helpful in the initial stages of assessment, before drug and oxygen administration. They fail to accurately determine the status or severity during reassessment of the patient after acute therapy.

Spirometry tests as mentioned earlier, such as the peak expiratory flow rate (PEF) and the FEV1, also often fail to perform as intended. While the PEF and FEV1 do measure maximum airflow through the breathing passages and are markedly reduced in asthmatic attacks, they require a forceful full expiration performed by the patient to produce an accurate and reliable result. Unfortunately, patients with an asthmatic attack are often too breathless to perform such forceful full expirations; and particularly, many children cannot perform those necessary efforts, therefore most spirometry results taken during the asthmatic attack are unreliable (Schifano et al., 2014). Low PEF or FEV1 readings therefore could mean either severe airflow obstruction or poor effort from the patient. This renders this useful test quite ineffective during acute asthmatic attacks. As patients improve with medications, PEF and FEV1 readings maybe become more accurate, but without the initial readings to compare with, this is often not of much help. It is this inability to continuously monitor the asthmatic patient throughout the asthmatic attack using the same monitoring tool, to assess their improvement or decline, and to see early trends of their rapidly changing status that drives us to consider other modes of monitoring asthmatic attacks. To do this, we have to go beyond currently known and practiced ways.

2.9 Is capnography an ideal mode of monitoring for asthmatic attacks?

The ideal monitoring mechanism for an acute asthmatic attack should be a measure of degree of airflow obstruction, which can be monitored passively, without any additional effort from the already-breathless patient, can be monitored continuously, without interfering with drug or oxygen administration; and can alert clinicians to early signs of deterioration. In this regard, capnographic waveform indices carry huge potential to deliver on all those fronts.

Capnography is the measurement of expired CO_2 in each expired breath. The amount of CO_2 is expired air is usually measured by a sampling device at the level of the mouth or nose. Measuring the capnograph using side-stream measurement mode is simple, akin to setting up a nasal oxygen cannula. Fig. 2.5 shows the typical capnographic waveform appears at the normal condition in a healthy patient. In fact, most capnograph cannulas do deliver O_2 as well as measure CO_2. The measurements are usually performed by absorption spectrometry many times per second, giving an up-to-date, and breath-by-breath reading of CO_2 levels. As a reminder, CO_2 in the expired air is a result of normal metabolism within all cells that burn glucose with oxygen to produce carbon dioxide CO_2. The CO_2 from the cells diffuses out into the circulation, where it is carried to the alveoli in the lungs

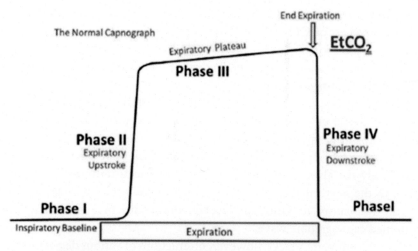

Figure 2.5 Typical capnographic waveform in healthy patient.

and diffuses out of the capillaries into the alveoli. (O_2 moves in the opposite direction; hence it is often referred to as O_2–CO_2 exchange, although in most part, it occurs independently of each other).

Within the breathing passages, CO_2 levels are highest in the alveoli and least in the air that we breathe in. Therefore, as we exhale, the initial parts of the exhaled breath will comprise of surrounding air, mixed with air in the breathing passages. This mixed air contains a variable but increasing amounts of CO_2 as air from the deeper and deeper air passages are exhaled and measured. This is seen in the capnograph as the sharply angled expiratory upstroke. At some point, the air from the alveoli, which has the highest amount of CO_2, is exhaled and measured. When the bronchioles are normally open and unobstructed, all the alveoli discharge their aim simultaneously, concurrent to exhalation. This means that alveolar air moves out of the millions of alveoli as a simultaneous wave, bringing with it the CO_2. This causes the measured CO_2 levels to rise sharply and then plateau off rapidly. This is seen as a horizontal plateau in the capnographic waveform (called the "Alveolar Plateau") signifying the expiration of alveolar air in a concurrent wave. The Alveolar Plateau usually ends with the highest reading ("end-tidal CO_2", Et CO_2), which is the reading at the end of expiration. The slope then drop back suddenly as the patient starts to inhale, and the reading now reflects the low levels of CO_2 in inspired environmental air.

End-tidal CO_2, Et CO_2, is commonly used in clinical situations, providing information about airway patency, giving estimates of blood

CO_2 levels, confirming placement of airway devices, and even monitoring the effectiveness of CPR efforts. However, the shape of the capnographic waveform also provides information about airflow. It is this specific characteristic of the capnographic waveform that we will use to measure airflow in the breathing passages.

In our understanding of airflow through the normal air passages, CO_2 from the alveoli exit through the air passages in a concurrent wave causing a steep expiratory upslope in the CO_2 waveform. The CO_2 levels then hit a plateau as the CO_2 from most of the alveoli exit at the same time, resulting in a relatively flat alveolar plateau. The alveolar plateau terminates with the onset of the next inhaled breath, which brings down the readings rapidly, on the inspiratory downstroke. Overall, this results in a normal square patterned capnographic wave.

During asthmatic attacks, however, obstructions of various degrees in the smaller breathing passages obstruct the flow of air exiting the alveoli, leading to a heterogeneity of CO_2 in expired air. In short, the varying degrees of obstruction in the many bronchioles cause the expiration of CO_2 to occur nonconcurrently and with a greater degree of mixing of air from all parts of the lungs. This will result in loss of sharply defined phases in the waveform, which causes changes in the *slope* of the capnograph; a decrease in the slope of Phase II (expiratory upstroke) and an increase in the slope of Phase III (alveolar plateau), as shown in Fig. 2.6 (Howe et al., 2011). This is due to the exhalation of CO_2 in a varied and mixed manner because of the various degrees of narrowing of the breathing passages in asthma.

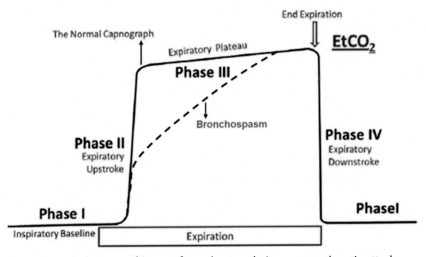

Figure 2.6 Capnographic waveform changes during acute asthmatic attack.

As the severity of disease worsens, and the degree of narrowing increases, the changes to the capnographic waveform also become more marked. The decrease in the slope of the expiratory upstroke (Phase II) and the increase in the slope of the alveolar plateau (Phase III) continue to develop until both phases appear indistinguishable from each other. A resulting "shark-fin" pattern emerges in the capnographic waveform. These waveform changes are particularly interesting because they indicate the degree of airflow obstruction, which directly reflects the narrowing of the air passages. Research has therefore postulated that these waveform changes are indicative of severity of asthmatic attacks. In various published studies to show this, the waveforms changes have been shown to correlate well with the assessed degree of severity (Nik Hisamuddin et al., 2009). They have also demonstrated good correlation with PEF and FEV1 spirometry readings. We can therefore contemplate the potential role of capnographic waveform analysis as a monitoring tool for asthmatic attacks.

As mentioned before, monitoring of the asthmatic attack is particularly difficult. Existing modes such as spirometry PEF or FEV1 readings are unreliable because they require full expiratory efforts by the patient, which is difficult when they are breathless during the attack. Here, the capnographic waveform has a particular advantage, as it does not require any particular effort from the patient. More importantly, the waveform changes can be seen live as it happens and the degree of change may reflect a worsening condition or improving response to treatment. The cannulas used to measure CO_2 can deliver O_2 at the same time and do not interfere with the delivery of inhaled or nebulized medications. So, in short, capnographic waveforms change in relation to the degree of narrowing of the breathing passages, both during the early stages of the asthmatic attack and during treatment. Combined with the fact that capnographic waveform monitoring is very much a noninvasive monitoring mode, which does not interfere with oxygen or drug administration, it is easy to spot the potential of capnographic waveform monitoring in asthma.

2.10 Capnographic waveform indices

However, visual interpretation of the capnographic waveform is inadequate to summarily make conclusions about severity. To be more specific, specific measurements of certain parameters in those waveforms are needed to be able to compare and contrast findings, correlate them with clinical findings, and estimate the reliability of those findings (Kean et al., 2010).

To further the regular and reliable use of capnographic waveforms in the management of asthma, it is essential to try to identify the best indices that reflect severity and to look at ways to reduce artifacts or noise that may skew the results.

The initial challenge would be to select the several most sensitive indices for analysis. In this regard, previous studies have looked at the slopes of Phase II and III and the angles created between those two slopes. We will look at and discuss more potential indices in the subsequent chapters. The next challenge would be to concurrently select suitable waveforms and filter out unsuitable waveforms. Among the reasons for unsuitable waveforms are inadequate breaths, incomplete expiration and gasping, and measurement artifacts. This may be seen as waveforms that are too small, do not reach the baseline or are too short in duration. The next challenge would be to accurately average out the waveform indices of each breath, without losing the sensitivity of the findings. This is a function of the filters that are applied to remove nonsuitable waveforms, artifacts, and noise. The better the filters are, the more sensitive the results. But filtering off too many waveforms may inadvertently create inaccurate results of its own. Lastly, as the waveform indices are developed, then a live-reporting mechanism of those indicators and their clinical significance will finally be able to aid the clinician in their assessment and improve their care for the patient. Being able to perform waveform analysis in which waveforms can be analyzed electronically and in live mode using various indices holds the key to using this tool in monitoring asthmatic patients. Algorithms that reduce artifacts and electronic noise, and immediately calculate indices hold the greatest promise for this potential new tool for monitoring asthmatic patients. Looking further forward, using artificial intelligence techniques to analyze the waveforms patterns may reveal even more useful indices, potentially creating the ability of such analytical devices to be better than humans in monitoring asthma.

The potential clinical use of automated capnographic waveform indices in the management of asthmatic attacks is not insignificant. If found reliable and useful, it will be a very cheap mode of monitoring asthmatic patients continuously, noninvasively, portably, and without interfering with drug or oxygen administration. It will be useful both before, during, and after medications and in all degrees of severity. A validated numeric indicator of airflow obstruction, which may be validly argued to be missing in our assessment as asthma as of today, will be a significant development in asthma monitoring and may well be a routine part of patient assessment soon. In addition, from this, using capnographic waveforms may also potentially aid the early diagnosis of asthma for patients. Especially in children who often find it difficult to describe symptoms of asthma, or in those who are unable to perform normal lung function tests, it may help reduce the high rates of undiagnosed asthma in the world. It will help clinicians decide on the need to continue therapy and facilitate admission/discharge decisions. Mostly it will potentially reduce the guesswork in assessment of asthmatic patients and render this safer for patients worldwide. Finally, the use of capnographic waveform indices may not restricted to asthma itself. The other causes of acute breathlessness in patients have varied etiologies, which variations may be reflected in different waveform morphologies. Ongoing research in this field of study will tell us more in the coming months and years.

References

Arora, G., & Bittner, V. (2015). Chest pain characteristics and gender in the early diagnosis of acute myocardial infarction. *Current Cardiology Reports, 17*(2), 1—5. https://doi.org/10.1007/s11886-014-0557-5

AsthmaUK. (2020). *Cough and wheeze*. AsthmaUK.

Atta, J. A., Avena, L. A., Borgiani, M. T., Fiorenza, R. F., Martins, M. A., & Medicina, F. De (2004). Patient and physician evaluation of the severity of acute asthma exacerbations. *Brazilian Journal of Medical and Biological Research, 37*(9), 1321—1330. https://doi.org/10.1590/S0100-879X2004000900005

Bacharier, L. B., Guilbert, T. W., Jartti, T., & Saglani, S. (2021). Which wheezing preschoolers should be treated for asthma? *Journal of Allergy and Clinical Immunology: In Practice, 9*(7), 2611—2618. https://doi.org/10.1016/j.jaip.2021.02.045

Baos, S., Calzada, D., Cremades-Jimeno, L., Sastre, J., Picado, C., Quiralte, J., Florido, F., Lahoz, C., & Cárdaba, B. (2018). Nonallergic asthma and its severity: Biomarkers for its discrimination in peripheral samples. *Frontiers in Immunology, 9*(JUN), 1—11. https://doi.org/10.3389/fimmu.2018.01416

Bijl-Hofland, I. D., Cloosterman, S. G. M., Folgering, H. T. M., Akkermans, R. P., & Van Schayck, C. P. (1999). Relation of the perception of airway obstruction to the severity of asthma. *Thorax, 54*(1), 15—19. https://doi.org/10.1136/thx.54.1.15

Boonpiyathad, T., Sözener, Z. C., Satitsuksanoa, P., & Akdis, C. A. (2019). Immunologic mechanisms in asthma. *Seminars in Immunology, 46*(October), 101333. https://doi.org/10.1016/j.smim.2019.101333

Bouazza, B., Hadj-Said, D., Pescatore, K. A., & Chahed, R. (2021). Are patients with asthma and chronic obstructive pulmonary disease preferred targets of COVID-19? *Tuberculosis and Respiratory Diseases, 84*(1), 22—34. https://doi.org/10.4046/TRD.2020.0101

Brown, A., Aristides, M., FitzGerald, P., Davey, P., Bhalla, S., & Kielhorn, A. (2002). Pcn19 examining preferences and timetrade-off utility for gemcitabine plus cisplatin in the treatment of bladder cancer:a survey using discrete choice conjoint analysis in the UK. *Value in Health, 5*(6), 543—544. https://doi.org/10.1016/s1098-3015(10)61435-0

Campbell, M. L. (2017). Dyspnea. *Critical Care Nursing Clinics of North America, 29*(4), 461—470. https://doi.org/10.1016/j.cnc.2017.08.006

Cartier, A. (2015). New causes of immunologic occupational asthma, 2012—2014. *Current Opinion in Allergy and Clinical Immunology, 15*(2), 117—123. https://doi.org/10.1097/ACI.0000000000000145

Choi, Y., Lee, Y., & Park, H. S. (2019). Neutrophil activation in occupational asthma. *Current Opinion in Allergy and Clinical Immunology, 19*(2), 81—85. https://doi.org/10.1097/ACI.0000000000000507

Côté, A., Turmel, J., & Boulet, L. P. (2018). Exercise and asthma. *Seminars in Respiratory and Critical Care Medicine, 39*(1), 19—28. https://doi.org/10.1055/s-0037-1606215

Cukic, V., Lovre, V., Dragisic, D., & Ustamujic, A. (2012). Asthma and chronic obstructive pulmonary disease (copd) and #8211; differences and similarities. *Materia Socio Medica, 24*(2), 100. https://doi.org/10.5455/msm.2012.24.100-105

Dharmage, S. C., Perret, J. L., & Custovic, A. (2019). Epidemiology of asthma in children and adults. *Frontiers in Pediatrics, 7*(JUN), 1—15. https://doi.org/10.3389/fped.2019.00246

Edmondstone, W. M. (1998). Chest pain and non respiratory symptoms in acute asthma. *Thorax, 53*(Suppl. 4), 413—414.

Fainardi, V., Santoro, A., & Caffarelli, C. (2020). Preschool wheezing: Trajectories and long-term treatment. *Frontiers in Pediatrics, 8*(May), 1—8. https://doi.org/10.3389/fped.2020.00240

Fang, L., Sun, Q., & Roth, M. (2020). Immunologic and non-immunologic mechanisms leading to airway remodeling in asthma. *International Journal of Molecular Sciences, 21*(3), 1—19. https://doi.org/10.3390/ijms21030757

Fergeson, J. E., Patel, S. S., & Lockey, R. F. (2017). Acute asthma, prognosis, and treatment. *Journal of Allergy and Clinical Immunology, 139*(2), 438—447. https://doi.org/10.1016/j.jaci.2016.06.054

Fernández-Nieto, M., Quirce, S., & Sastre, J. (2006). Occupational asthma in industry. *Allergologia et Immunopathologia, 34*(5), 212—223. https://doi.org/10.1157/13094029

Friedlander, J. L., Baxi, S., & Phipatanakul, W. (2014). Asthma and allergens. *Clinical Asthma: Theory and Practice,* 93—99. https://doi.org/10.1201/b16468

Halpin, D. M. G. (2020). What is asthma chronic obstructive pulmonary disease overlap? *Clinics in Chest Medicine, 41*(3), 395—403. https://doi.org/10.1016/j.ccm.2020.06.006

Holgate, S. T., & Thomas, M. (2017). Asthma. In *Middleton's allergy essentials* (1st ed., pp. 151—204). Elsevier Inc. https://doi.org/10.1016/B978-0-323-37579-5.00007-6

Howe, T. A., Jaalam, K., Ahmad, R., Sheng, C. K., & Nik Ab Rahman, N. H. (2011). The use of end-tidal capnography to monitor non-intubated patients presenting with acute exacerbation of asthma in the emergency department. *Journal of Emergency Medicine, 41*(6), 581—589. https://doi.org/10.1016/j.jemermed.2008.10.017

Ismail, H. I. H. M., Ibrahim, H. M., Phak, N. H., & Thomas, T. (2019). Paediatric paediatric. In *Paediatric protocols for Malaysian hospitals* (4th ed., pp. 1—596).

Jarvis, J., Seed, M. J., Elton, R. A., Sawyer, L., & Agius, R. M. (2005). Relationship between chemical structure and the occupational asthma hazard of low molecular weight organic compounds. *Occupational and Environmental Medicine, 62*(4), 243—250. https://doi.org/10.1136/oem.2004.016402

Kean, T. T., Teo, A. H., & Malarvili, M. B. (2010). Feature extraction of capnogram for asthmatic patient. In , *Vol 2. 2010 2nd international conference on computer engineering and applications* (pp. 251—255). ICCEA. https://doi.org/10.1109/ICCEA.2010.286

Kuprys-Lipinska, I., Kolacinska-Flont, M., & Kuna, P. (2020). New approach to intermittent and mild asthma therapy: Evolution or revolution in the GINA guidelines? *Clinical and Translational Allergy, 10*(1), 1—14. https://doi.org/10.1186/s13601-020-00316-z

Kwah, J. H., & Peters, A. T. (2019). Asthma in adults: Principles of treatment. *Allergy and Asthma Proceedings, 40*(6), 396—402. https://doi.org/10.2500/aap.2019.40.4256

Kwong, C. G., & Bacharier, L. B. (2019). Phenotypes of wheezing and asthma in preschool children. *Current Opinion in Allergy and Clinical Immunology, 19*(2), 148—153. https://doi.org/10.1097/ACI.0000000000000516

Lang, J. E. (2019). The impact of exercise on asthma. *Current Opinion in Allergy and Clinical Immunology, 19*(2), 118−125. https://doi.org/10.1097/ACI.0000000000000510

Le Souëf, P. (2018). Viral infections in wheezing disorders. *European Respiratory Review, 27*(147). https://doi.org/10.1183/16000617.0133-2017

Maike Grotheer, A., Schulz, B., Respirationstrakt, S., Bronchialerkrankung, Chronische, & Grotheer Medizinische Kleintierklinik, M. (2019). Felines Asthma und chronische Bronchitis-Übersicht zu Diagnostik und Therapie Feline asthma and chronic bronchitis-an overview of diagnostics and therapy. *Felines Asthma Und … Tierarztl Prax Ausg K Kleintiere Heimtiere, 47*, 175−188. https://doi.org/10.1055/a-0917-6245

Martinez, F. D., & Vercelli, D. (2013). Asthma. *The Lancet, 382*(9901), 1360−1372. https://doi.org/10.1016/S0140-6736(13)61536-6

McCallion, P., & De Soyza, A. (2017). Cough and bronchiectasis. *Pulmonary Pharmacology and Therapeutics, 47*, 77−83. https://doi.org/10.1016/j.pupt.2017.04.010

MD, J. W. M. (2015). Asthma: Definitions and pathophysiology. *International Forum of Allergy and Rhinology, 5*(September), S2−S6. https://doi.org/10.1002/alr.21609

Miraglia del Giudice, M., Allegorico, A., Parisi, G., Galdo, F., Alterio, E., Coronella, A., Campana, G., Indolfi, C., Valenti, N., Di Prisco, S., Caggiano, S., & Maiello, N. (2014). Risk factors for asthma. *Italian Journal of Pediatrics, 40*(1), 1−2. https://doi.org/10.1186/1824-7288-40-S1-A77

Nema, S. (2004). Robbins basic pathology - (2003). *Medical Journal Armed Forces India, 60*(1), 92. https://doi.org/10.1016/s0377-1237(04)80179-5

de Nijs, S. B., Venekamp, L. N., & Bel, E. H. (2013). Adult-onset asthma: Is it really different? *European Respiratory Review, 22*(127), 44−52. https://doi.org/10.1183/09059180.00007112

Nik Hisamuddin, N. A. R., Rashidi, A., Chew, K. S., Kamaruddin, J., Idzwan, Z., & Teo, A. H. (2009). Correlations between capnographic waveforms and peak flow meter measurement in emergency department management of asthma. *International Journal of Emergency Medicine, 2*(2), 83−89. https://doi.org/10.1007/s12245-009-0088-9

O'Donnell, D. E., Milne, K. M., James, M. D., de Torres, J. P., & Neder, J. A. (2020). Dyspnea in COPD: New mechanistic insights and management implications. *Advances in Therapy, 37*(1), 41−60. https://doi.org/10.1007/s12325-019-01128-9

Ogawa, H., Fujimura, M., Ohkura, N., & Makimura, K. (2014). Atopic cough and fungal allergy. *Journal of Thoracic Disease, 6*(10), S689−S698. https://doi.org/10.3978/j.issn.2072-1439.2014.09.25

Ortiz, G., & Sanders, D. H. (2012). Adult asthma. *Journal of Asthma & Allergy Educators, 3*(3), 129−131. https://doi.org/10.1177/2150129712448612

Page, B. (2012). Severe asthma. *JEMS: A Journal of Emergency Medical Services*, 16−19. https://doi.org/10.3238/arztebl.2014.0847. Suppl.

Pakkasela, J., Ilmarinen, P., Honkamäki, J., Tuomisto, L. E., Andersén, H., Piirilä, P., Hisinger-Mölkänen, H., Sovijärvi, A., Backman, H., Lundbäck, B., Rönmark, E., Kankaanranta, H., & Lehtimäki, L. (2020). Age-specific incidence of allergic and non-allergic asthma. *BMC Pulmonary Medicine, 20*(1), 1−9. https://doi.org/10.1186/s12890-019-1040-2

Patadia, M. O., Murrill, L. L., & Corey, J. (2014). Asthma. Symptoms and presentation. *Otolaryngologic Clinics of North America, 47*(1), 23−32. https://doi.org/10.1016/j.otc.2013.10.001

Peters, S. P. (2014). Asthma phenotypes: Nonallergic (intrinsic) asthma. *Journal of Allergy and Clinical Immunology: In Practice, 2*(6), 650−652. https://doi.org/10.1016/j.jaip.2014.09.006

Prevention, C. for D. C. (2020). *Common asthma triggers.* CDC.

Quirce, S., & Sastre, J. (2019). Occupational asthma: Clinical phenotypes, biomarkers, and management. *Current Opinion in Pulmonary Medicine, 25*(1), 59–63. https://doi.org/10.1097/MCP.0000000000000535

Quirt, J., Hildebrand, K. J., Mazza, J., Noya, F., & Kim, H. (2018). Asthma. *Allergy, Asthma and Clinical Immunology, 14*(Suppl. 2). https://doi.org/10.1186/s13223-018-0279-0

Rodriguez Bauza, D. E., & Silveyra, P. (2021). Asthma, atopy, and exercise: Sex differences in exercise-induced bronchoconstriction. *Experimental Biology and Medicine, 246*(12), 1400–1409. https://doi.org/10.1177/15353702211003858

Saadeh, C. K. (2020). Status asthmaticus. *Medscape*, 1–31, 2129484.

Schifano, E. D., Hollenbach, J. P., & Cloutier, M. M. (2014). Mismatch between asthma symptoms and spirometry: Implications for managing asthma in children. *Journal of Pediatrics, 165*(5), 997–1002. https://doi.org/10.1016/j.jpeds.2014.07.026

Tarlo, S. M., & Lemiere, C. (2014). Occupational Health. *The New England Journal of Medicine, 37*, 640–649. https://doi.org/10.1016/j.ssci.2022.105707

Tiotiu, A. I., Novakova, S., Labor, M., Emelyanov, A., Mihaicuta, S., Novakova, P., & Nedeva, D. (2020). Progress in occupational asthma. *International Journal of Environmental Research and Public Health, 17*(12), 1–19. https://doi.org/10.3390/ijerph17124553

Uddin, M., Aiello, A. E., Wildman, D. E., Koenen, K. C., Pawelec, G., De Los Santos, R., Goldmann, E., & Galea, S. (2010). Epigenetic and immune function profiles associated with posttraumatic stress disorder. *Proceedings of the National Academy of Sciences of the United States of America, 107*(20), 9470–9475. https://doi.org/10.1073/pnas.0910794107

Vigeland, C. L., Hughes, A. H., & Horton, M. R. (2017). Etiology and treatment of cough in idiopathic pulmonary fibrosis. *Respiratory Medicine, 123*, 98–104. https://doi.org/10.1016/j.rmed.2016.12.016

WHO. (2021). *Asthma*. WHO.

CHAPTER THREE

Current tools for assessment of asthma

Respiratory diseases including asthma are very common chronic mortality diseases, which affect all age groups. These diseases have been known to humans for many decades, which helped researchers to come up with devices capable of diagnosing these diseases. Throughout previous decades, these devices have undergone significant changes resultantly improving their performances. Improved performance has consequently enhanced the quality of life substantially. There are many clinical devices, which are currently used to measure different parameters to perform pulmonary diagnosis. Existing devices mainly measure pressure, flow, and volume and gas concentration to do diagnosis. These devices either use one or many of the abovementioned parameters in combination to perform in detail or advanced diagnosis. Carbon dioxide (CO_2) is naturally produced in our body, which is removed primarily through ventilation when breath is exhaled. If a person is suffering from breathing problems, the amount of CO_2 coming out of his/her body will be reduced resultantly increasing the level of CO_2 in the body. Increased CO_2 level can cause respiratory depression or even respiratory failure. Some of many available clinical devices being used for diagnosis are discussed below along with their benefits as well as limitations.

3.1 Diagnosis of asthma: common techniques and limitations

Asthma is a common respiratory disease, and it affects all age groups. Asthmatic patients mostly experience struggles while breathing because of the resistance to airflow that causes the person to exhale with extra effort (Guyton & Hall, 1986). Diagnosis of asthma can be complicated due to several clinical and molecular heterogeneity of this disease (example: atopic vs. nonatopic, obesity-associated asthma), and this affects response to the therapy. Asthma is manifested by recurrent episodes of shortness of breath, chest tightest, cough, and wheeze (Sokol et al., 2015). The diagnosis of asthma should not only rely on symptoms as they are not specific and

Systems and Signal Processing of Capnography as a Diagnostic Tool for Asthma Assessment
ISBN: 978-0-323-85747-5
https://doi.org/10.1016/B978-0-323-85747-5.00006-1

may improve due to the treatment (Global Initiative for Asthma, 2021). The diagnosis based on physical examination involves the assessment of the conditions of the upper respiratory tract, the chest, and the skin (Education & Program, 2007). Both physical examination and medical history are not reliable enough while diagnosing asthma. There is no stand-alone diagnostic test to examine asthma in a particular patient. Thus, affective diagnosis of asthma should be based on patient's medical history, physical examination, and spirometry as an objective assessment tool. Additional tests are also recommended in the diagnosis of asthma as necessary to preclude alternative diagnoses (National Heart & Institute, 2007). In patients with asthma, there is a great variability in signs and symptoms over time, and this variability is different from patients to patient. Therefore, clinical judgment is of importance while assessing asthma.

3.1.1 Medical history documentation

Asthma is a symptomatic disease. The key indicators for considering a diagnosis of asthma include wheezing; recurrent episodes of cough, chest tightness and breathing difficulties; precipitating factors; and sleep disturbance due to symptoms that occur or worsens at night (National Heart & Institute, 2007). Cough is the predominant symptom of asthma. However, if the patient is presenting only cough, other tests such as pulmonary function testing must be carried out to assess the presence of airway obstruction. On the other hands, the problem rises for young children having only cough as they find it difficult to perform pulmonary function testing to confirm the presence of asthma (Castro & Kraft, 2008). Asthma symptoms may be precipitated by different factors such as exercise, tobacco smoke, viral infection, changes in the weather, strong emotional expression, and exposure to airborne allergens and dusts (Castro & Kraft, 2008; National Heart & Institute, 2007). When multiple symptoms are present, the probability of asthma is increased. However, spirometry should be conducted to establish a diagnosis. The symptoms that are probably to be caused by asthma can be identified while conducting a thorough medical history. In the NHLBI, 2007, the expert panel has recommended a list of items that are to be addressed in a detailed medical history of the new patient who is known or suspected to have asthma (National Heart & Institute, 2007). Suggested items for medical history are summarized in Fig. 3.1.

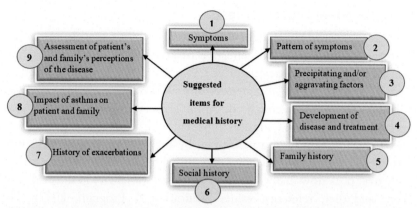

Figure 3.1 Suggested items for medical history for a diagnosis of asthma in a new patient of known or thought to have asthma (National Heart & Institute, 2007). *This list does not represent a standardized assessment or diagnostic instrument. The validity and reliability of this list have not been assessed. *From National Heart, Lung, and Blood Institute (2007).*

3.1.2 Physical examination

Physical examination for asthma deals with the evaluation of the status of the upper respiratory tract, chest, and skin. Abnormality in the upper respiratory tract can be reflected by increased nasal discharge, swollen mucosa, and sometimes nasal polyps can occur. Chest examination involves evaluation of the thoracic hyperexpansion, chest deformity, and the sounds of wheezing during tidal breathing. Assessment of the skin examines the presence of atopic dermatitis (Education & Program, 2007). The absence of physical findings does not rule out asthma, as the disease has a variable nature, and signs of airflow obstruction are not mostly present between asthma episodes. Furthermore, physical examination is usually normal unless the disease is at the severe stage, or the examination is conducted during exacerbation of the disease (Kavanagh et al., 2019). Expiratory wheezing examined on auscultation is the most frequent abnormality in physical examination; however, wheezing may only be heard when the exhalation is done forcefully. Wheezing may be absent when severe exacerbation is at the severe stage as a result of declined airflow (so called "silent chest") (Global Initiative for Asthma, 2021). In addition, wheezing with characteristics similar to asthma can be identified in different diseases (Castro & Kraft, 2008) such as chronic obstructive pulmonary disease (COPD), respiratory infections, tracheomalacia, or inhaled foreign body.

3.1.3 Pulmonary function tests

Assessment of variable expiratory airflow limitation is necessary to confirm the diagnosis and to provide effective treatment. Unconfirmed diagnosis can lead to unnecessary treatment or leaving behind other important diagnoses. If possible, variable expiratory airflow should be confirmed before initiating treatment for asthma (Global Initiative for Asthma, 2021). Expiratory airflow limitation is common in patient with asthma, and it varies with time. Lung function testing performed by using spirometer or peak flow meter helps to track the variability of airflow limitation. Spirometry is preferable over peak flow meter due to the extensive variability in the peak flow meters and their reference values (Education & Program, 2007). The following subsections discuss the traditional diagnostic tools for asthmatic conditions and their limitations.

3.1.3.1 Spirometry

Spirometry is the most frequent pulmonary function test that is performed to objectively assess the lung condition and monitor respiratory airway diseases (Graham et al., 2019). Spirometry indicates how much air an individual can forcefully and completely breath out after quick and maximum inhalation. In asthmatic patients, a well-performed spirometry indicates the level of airway obstruction, severity stage, and reversibility of the disease (van den Wijngaart et al., 2015). Perception of patients regarding airflow obstruction is highly different, and the severity of the disease revealed by spirometry is sometimes very high as compared to that would have been estimated based on the history and physical examination. Thus, consistent monitoring of pulmonary function can help to assist asthmatic patients who underperceive their symptoms until the airflow limitation reaches at the severe stage (National Heart & Institute, 2007). With spirometry, exacerbations can also be examined as exacerbations of asthma are associated with reduction in the expiratory airflow. Different spirometry parameters and their clinical interpretations are discussed in the next section.

3.1.3.1.1 *Spirometry parameters and their interpretation*

The most commonly spirometry parameters that are measured at clinical practice are the FEV_1, forced vital capacity (FVC), and their ratio (FEV_1/FVC). The FEV_1 is the most commonly obtained and reproducible parameter, and it corresponds to the amount of air breathed out in the first second during forced expiration following fully inhalation (Hyatt et al., 2014). FVC is the total volume of air exhaled forcibly in one breath after the utmost inhalation (Dempsey & Scanlon, 2018). The majority of patients can provide acceptable and repeatable results when a well-trained technician is available to provide adequate coaching while performing spirometry test. For healthy young children, it takes few seconds to complete exhalation of the whole vital capacity compared to older subjects. In adult patients with airflow obstruction, a maximal expiratory effort can be completed for more than 12 or 15 seconds and thus, some patients find it prolonged, and they feel uncomfortable to complete the maneuver. Therefore, the measurement of the FEV_6 has taken place as an alternative for measurement of FVC in adults (National Heart & Institute, 2007). However, the FEV_1/FEV_6 should be used with caution to avoid the risk of overdiagnosis of airflow obstruction in the elders (Chung et al., 2016).

Spirometry measurements help to determine whether there is airflow obstruction, its severity, and whether the obstruction is reversible over the short term. Based on spirometry results, respiratory airways may be in normal condition or have restrictive, or obstructive, or mixed pattern. Normal spirometry is reflected by the values of FVC, FEV_1, and FEV_1/FVC that fall within the normal range (Haynes, 2018). However, having spirometry values that are in the normal range cannot totally rule out the lung disease. In this, many patients with asthma mostly manifest normal lung function (Haynes, 2018). In the restrictive airways, the FVC is decreased but the FEV_1 is normal or decreased. Thus, the FEV_1/FVC ratio is normal or increased (Jat, 2013; National Heart & Institute, 2007). On the other hand, in obstructive pattern, the airway resistance is increased, and it restricts the patient to rapidly exhale the air. The FVC and FEV1 both are smaller compared to the normal range, and the FEV_1/FVC ratio is reduced (Barreiro & Perillo, 2004). Mixed ventilatory defect means that the patient has both restriction and obstruction in the respiratory airways. This type of defect cannot be diagnosed using only spirometry, as vital capacity reduction can be caused by the airway obstruction without involving the airway restriction (Levy et al., 2009).

Spirometry maneuvers can graphically be represented as volume—time curve or as flow-volume loop. Volume—time curve plots the volume in liters versus time (in second). While flow-volume loop displays flow (in liters/second) vertically on the y-axis and volume (in liters) horizontally on the x-axis. Exhalation phase is indicated by the flows located above the intercept of the horizontal axis, whereas inhalation phase corresponds to the flows below the intercept of the horizontal axis (Haynes, 2018). Flow-volume loop can help to identify common spirometry errors. In a healthy subject, the FEV_1 is higher, and the flow-volume loop is tall and wide as it can be observed on Fig. 3.2A and B. In patient with obstructive disease, the FEV1 is reduced, and the flow-volume loop is deformed in shape. The expiratory loop is concave and peak flow and mid-expiratory flows are decreased (Fig. 3.2C and D).

A value of forced expiratory flow over the middle half of the FVC ($FEF_{25-75}\%$) is another spirometry derived flow measurement that may indicate the presence of airflow obstruction in the airways (Jat, 2013). The $FEF_{25-75}\%$ is thought to be more specific to small airway patency, especially when the FEV_1 is normal (Johns et al., 2014). The reduction in the FEF_{25-75} might be an early marker for the FEV_1 impairment, thus revealing

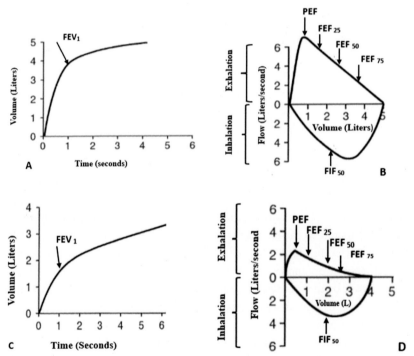

Figure 3.2 Spirometry measurements during forced expiration. (A) and (B) correspond to the volume—time curve and the flow-volume loop of a healthy subject, whereas (C) and (D) are the volume—time curve, and the flow-volume loop of a patient with airway obstruction. FEV_1 is the forced expiratory volume in the first second; PEF is the peak expiratory flow; FEF is the forced expiratory flow; FEF_{25}, FEF_{50}, and FEF_{75} are the mid-expiratory flows after 25%, 50%, and 75% of the FVC has been exhaled, respectively. FIF_{50}, forced inspiratory flow measured at the same volume as FEF_{50} (Hyatt et al., 2014).

early disease and poor prognosis. In a cross-sectional study carried out on 602 asthmatic patients, the low value of FEF_{25-75} was suggested to be an indicator for uncontrolled or partially controlled asthma. However, more studies are required to confirm the relevance of this parameter as a prognostic index in common practice (Ciprandi et al., 2018). Spirometry parameters are among components considered while classifying asthma severity levels in young children, youths, and adults. Tables 3.1 and 3.2 show the ranges of FEV_1 and FEV_1/FVC values in patient with intermittent and those with persistent asthma (Education & Program, 2007).

The most frequently numeric values use to interpret spirometry results are FEV_1, FVC, and the ratio FEV_1/FVC (Jat, 2013). Airflow obstruction is characterized by a decrease in both the FEV_1 and the FEV_1/FVC values in respect to reference or predicted values. According to the NHLBI asthma guidelines, patients with intermittent and mild asthma have FEV_1 values of more than 80% in all age groups. The FEV_1 varies between 60% and 80% in patients with moderate persistent asthma, whereas those with severe persistent asthma have FEV_1 values of less than 60% of predicted value. The FEV_1 primarily measures flow through the mid- to large-sized airways. The low

Table 3.1 The range of values of spirometry parameters in young children (5—11 years old) with asthma (Education & Program, 2007).

| | | Asthma severity level | | |
| | | Persistent | | |
Spirometry parameter	Intermittent	Mild	Moderate	Severe
FEV_1 (predicted) (in %)	>80	>80	60—80	<60
FEV_1/FVC (in %)	>85	>80	75—80	<75

Table 3.2 Spirometry parameters in asthmatic subjects with ≥12 years old (Education & Program, 2007).

| | | Asthma severity level | | |
| | | Persistent | | |
Spirometry parameter	Intermittent	Mild	Moderate	Severe
FEV_1 (%)	>80 predicted	>80 predicted	60—80 (predicted)	<60 predicted
FEV_1/FVC	Normal	Normal	FEV_1/FVC reduced 5%	FEV_1/FVC reduced >5%

FEV_1 can reflect the present airflow obstruction, and it is also a good predictor for future exacerbations (Castro & Kraft, 2008). An increase or reduction in FEV_1 of greater than 12% and greater than 200 mL from baseline is accepted as being consistent with asthma, in adults with respiratory symptoms typical of asthma (Global Initiative for Asthma, 2021). The FEV_1/FVC is the most sensitive and specific indicator for identifying airway obstruction (Jat, 2013). The FEV_1/FVC ratio is the highest in young children with intermittent asthma, and it is decreased with different stages of persistent asthma (Table 3.1). The normal value of the FEV_1/FVC is 85% for children, it varies between 75% and 80% for adults, while for elders, the FEV_1/FVC is 70% (Education and Program, 2007). For adults, lung function decreases as their ages increase and the higher declination can be identified in subjects who have asthma or those who are exposed to active smoking. On contrary, lung function increases for healthy children as they grow older until the optimal lung function, which is achieved by the age of 20 years for majority of individuals. Due to asthma, children may experience reductions in the lung growth compared to children with normal respiratory airways. Measurement of postbronchodilator FEV_1 can help to monitor lung growth patterns over time and the reduction in the obtained readings may be an indication that asthma control is progressively getting worse (National Heart & Institute, 2007).

3.1.3.1.2 Types of spirometers

As technology evolves, spirometers come in different designs. For example, Fig. 3.3 shows a wet spirometer consisting of an upright, water-filled cylinder and a bell which is a smaller chamber inverted inside the first and partially immersed in water. The counterweight and marker (indicator) are attached to the inverted chamber (Hazari & Costa, 2010). A water-filled cylinder has a breathing tube connected to it, and when the person breathes, the air passes through the tube into the inverted chamber, causing it to move in upward and downward directions. The speed of the bell depends on how fast the subject is breathing. The movement of the bell simultaneously moves the marker along a moving scale to record the lung volume measurements on a paper attached to the rotating drum a Kymograph.

The movement of the bell is proportional to the tidal volume. In most applications, the capacity of the bell ranges from 6 to 8 L. The normal spirometer has only the ability of responding fully to slow respiratory rates and not to rapid breathing that are occasionally experienced after anesthesia (Khandpur, 2005). Earlier spirometers have played an important role in

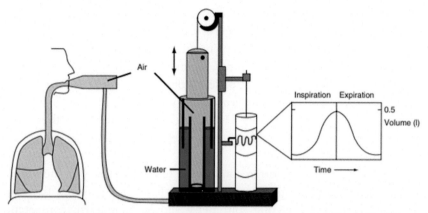

Figure 3.3 Illustration of a wet spirometer consisting of a water-filled chamber, an inverted bell, a counterweight and a marker to record lung volume measurements (Hazari & Costa, 2010).

disseminating spirometry measurement and application. These spirometers are still in use in some primary cares, despite their limitations that they are heavy and usually not portable, it is difficult to clean and to disinfect them and their calibration is difficult as well (Johns et al., 2014). They can develop the leakages, which can lead to the improper collection of accurate volume (Wanger, 2012). With these spirometers, no flow-volume curve is produced, as spirometry variables are calculated manually (Johns et al., 2014). This is time-consuming, and it can result in the measurement errors.

Most recent generation of spirometers displays real-time graphs usually spirogram and flow-volume curve. They have built-in printers, and they can be connected to the computer to visualize the results (Jat, 2013). In addition, modern spirometers have infection control, and no daily calibration is required. The newly spirometer to be selected for practical purpose should meet the recommendations developed jointly by American Thoracic Society/European Respiratory Society (ATS/ERS) recommendations for spirometry. Accurate and repeatable spirometry requires the exhaled gas to be regulated to body temperature and humidity (BTPS) (Levy et al., 2009).

3.1.3.1.3 Limitations of spirometry

Spirometry is a safe and useful diagnostic tool to evaluate the lung function. However, the accuracy of the test depends on various factors including the willingness of the patient to perform the test, and the ability to understand and to follow a set of instructions given during the maneuver. Therefore, children of less than 5 years old, elders, and the patients suffering from chest pain find it difficult to perform spirometry (Jerzyńska et al., 2015). Substantial physical effort and interest of the patient are required to obtain accurate and repeatable spirometry. Different incidences that can lead to the poor performance include hesitation in test, submaximal effort at start of exhalation, premature finish and restart, patient stops too soon before the maneuver is completed, and coughing during the first second of the test, which can lead to the reduction in the FEV_1 (Hyatt et al., 2014). Once any of the above conditions occurs, the patient is required to repeat the test so that acceptable spirometry can be achieved.

The skill of the technician to conduct the test and to interpret the results has also impact on the spirometry outcome. It is recommended that the personnel who is conducting spirometry should be trained, competent and experienced in order to adequately coach the patient to perform the test and then to evaluate the quality of obtained graphs immediately after the maneuver is completed. Interpretation of the results, right after the test will help the patient to not be recalled later once any problem is identified (Levy et al., 2009). A great number of patients with lung diseases attend primary cares and there are treated there. However, in some primary cares, spirometry is underused, especially in the patients that are initially diagnosed with asthma. This may probably result from the lack of information related to the current guidelines to perform spirometry (Sokol et al., 2015). Underutilization of spirometry can also be caused by the insufficiency of spirometers in the clinical settings as well as the lack of appropriate training and comfort to perform the test and to interpret the results (Thompson & Jaffe, 2005). There is still a lack of quality assurance of the spirometry performed in the primary cares.

3.1.3.2 Peak flow meter

Peak flow meter is a small, cost-effective, portable, and easy-to-use device that measures the peak flow of the air that a person can breathe out forcefully after taking deep inhalation (Adeniyi & Erhabor, 2011). During an asthma episode, airways of the lungs are usually narrowed, and the peak flow meter can help to objectively assess the respiratory airways condition even before the manifestation of symptoms and to determine the degree of obstruction associated to them. Furthermore, peak flow monitoring in patients diagnosed with asthma, is useful to evaluate the course of the disease and how the patient is responding to treatment (Nazir et al., 2005). And thus, timely intervention can be provided, which subsequently limits unnecessary hospital admission. During asthma exacerbation, reduction in the expiratory flow can be quantified by peak expiratory flow measurement. In adults presenting respiratory symptoms typical of asthma, a change of at least 20% in the PEF, is accepted as being consistent with asthma (Global Initiative for Asthma, 2021). Peak flow readings should be checked regularly as they change over time. Monitoring peak flow measurement periodically on daily basis, may also be helpful to identify how bronchospasm is linked to the environmental or occupational exposures (National Heart & Institute, 2007).

Peak flow meter can be used at home environment to help patients to monitor themselves responses to treatment. Thus, it is important for asthmatic patients to know how to use peak flow meter. There are several designs of peak flow meters available on the market. However, their working principle is the same. While performing the measurement, the patient is instructed to be in the sitting or standing up position (DeVrieze et al., 2017). The maneuver starts by placing the indicator to the bottom of the numbered scale of the peak flow meter device. The patient is requested to fully fill the lungs by taking a deep breath and then to place the mouthpiece in the mouth. Thereafter, the patient blows out in a single blow as hard and fast as possible. The test is repeated three times and the highest value is recorded in patient's asthma dairy. Of importance, when the patient coughs or make a mistake, obtained value is not counted and the test have to be repeated (National Heart & Institute, 2007). Variability in the daily PEF is computed from two readings recorded every day. In healthy adults, the upper 95% confidence limit of diurnal variability from twice daily measurements is 9%,whereas, in healthy children it is 12.3%. Consequently, diurnal variability above 10% for adults and greater than 13% for children is regarded as huge variation in the PEF and it is an indicative that asthma control is not good and there is a high possibility of exacerbation (Global Initiative for Asthma, 2021).

3.1.3.2.1 Peak flow meter readings and their significance

For adequate monitoring of asthma using peak flow meter, it is very crucial to identify patient's personal best peak flow number as this number helps to determine the treatment plan appropriate to a given patient. Patient's personal best peak flow is the upmost peak flow reading the patient can achieve for at least two to three times in a day for a period of 2 weeks when he/she is feeling good and not manifesting any asthma symptom (i.e., asthma is under good control) (Adeniyi & Erhabor, 2011). It is considered as a standard value against which other readings are measured. By examining the patient's personal best when no symptoms are present, changes in asthma control can be identified. Asthma is under good control, when the PEF is consistently at a high level (Adeniyi & Erhabor, 2011). The personal best flow changes from patient to patient and thus, every asthmatic patient has a unique personal best flow number. When measuring PFR, the patient is suggested to use the same meter each time. Because, when a different meter is used, there may be a difference of approximately 20% in the readings (Global Initiative for Asthma, 2021). Despite the effort and technique required while measuring the PFR, the patients can still know their personal best flows when assisted by physicians to instruct and demonstrate them how to properly use the device (National Heart & Institute, 2007).

Typical peak flow meters have gauge markers that show three zones (Fig. 3.4). For the interpretation of peak flow scores, the colors of a traditional traffic light, green, yellow, and red, are used to indicate three zones (DeVrieze et al., 2017).

Green zone is the first region, and it is usually set at 80%—100% of patient's personal best or of predicted average. It is considered as a safe region, and it can be an indication that the patient is improving well with the help of prescribed medication. In this condition, the patient can perform normal activities and no difficulties to breath while sleeping (Adeniyi & Erhabor, 2011). Yellow zone indicates that the patient is within 50%—79% of

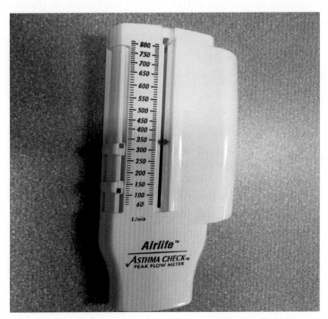

Figure 3.4 A typical peak flow meter for asthma monitoring (DeVrieze et al., 2017).

personal best and this may indicate that the patient must take measures to control asthma and the medications must be regulated as prescribed by the healthcare provider. This may also reflect the narrowing of the patient's respiratory airways and thus, the requirement for the supplemental daily medicines. Red is the last region, and it indicates that the peak flow reading is less than 50% of the patient's personal best. The patient may experience different symptoms such as cough, dyspnoea and distracted sleep. This is an emergency medical condition, and the patient is advised to directly contact the doctor or to go to the hospital emergency room (South Bay Allergy and Asthma, 2022). The PEF lower than 40% is a good predictor of the need for hospitalization (Education & Program, 2007).

3.1.3.2.2 Factors affecting normal peak flow value
In the subject with normal airways, peak flow readings are higher, and the values decrease when the airways become narrow. The normal peak low rates vary according to age, height, and gender (Adeniyi & Erhabor, 2011).

Peak flow readings are lower in women compared to men. For women, the PFR reaches the highest value approximately at the age of 30 years old and for men the utmost value is attained at 35 years old. In fact, lungs of women are smaller as compared to that of men of a given height (Hyatt et al., 2014). The peak flow readings gradually decline with age. This can be justified by the fact that the lung elasticity is reduced as the person is aging and subsequentially, the airways are smaller and the airflows are lower as well (Hyatt et al., 2014). The maximum values vary also depending on the height of the subject. The body height has the effect on the size of the lungs and its airways. A healthy taller subject has larger lung and airways compared to the shorter one of a given gender and age. As a result, taller subject has higher maximum expiratory flow (Hyatt et al., 2014). However, the influence of the height on the PEFR is lower compared to that of age and gender (Thaslima & Gayatri, 2020). In addition, body height should not be a parameter to be considered while estimating normal values for subjects with kyphoscoliosis. The reason is that in this category of subjects, the decreased height will lead to the overall underestimation of the normal lung volume and flows (Hyatt et al., 2014).

3.1.3.2.3 Limitations of peak flow meter
Peak flow measurement is highly effort dependent. The patient is required to forcefully blow out as fast as possible after maximum inhalation (Adeniyi & Erhabor, 2011). The test is repeated three times and the highest result of

the three readings is the actual peak flow rate. Thus, it is difficult for some patients to perform the test, particularly those who are in the emergency department as they may experience distress while blowing into the peak flow meter (Hisamuddin et al., 2009). Besides, there is a lack of repeatability of peak flow measurement as well as its correlation with other measures of airway function and consequently can lead to the underestimation of severity of airway obstruction in patient with asthma (Castro & Kraft, 2008). Due to a wide variability in the reference values of the predicted peak expiratory flow, peak flow meters are recommended to be utilized as monitoring rather than diagnostic tools (National Heart & Institute, 2007). In fact, each brand of peak flow meter is required to have specific reference values. However, currently majority of brands lack those specific reference values (National Heart & Institute, 2007).

3.1.4 Other tests for the assessment of asthma

3.1.4.1 Bronchial challenge testing

Bronchial challenge testing is useful in ruling out or confirming a suspected diagnosis of asthma, especially in patients with normal or near-normal spirometry results. The diagnosis of asthma is commonly done based on clinical history, physical examination, and spirometry as a confirmatory test for reversible airflow obstruction. When the patient manifests symptoms that are consistent to asthma and spirometry is not conclusive, then a bronchial challenge testing can be performed, because the results can reflect an increase or decrease probability of asthma (Coates et al., 2017). Bronchial challenge testing is conducted to assess airway hyperresponsiveness (AHR), defined as exaggerated response to direct or indirect stimuli that cause narrowing of the respiratory airways. Airway hyperresponsiveness is a clinical feature of asthma, however it may also occur in other diseases that cause inflammation or obstruction of respiratory airways (Coates et al., 2017). Bronchial challenge testing is classified as direct challenge with methacholine or histamine used as stimuli or indirect challenge where the mannitol or exercise are used as stimuli (Kaplan et al., 2009). Methacholine challenge is the most frequently performed test compared to the exercise challenge as it is easier to perform and it has higher sensitivity for a diagnosis of asthma, however specificity is less. Exercise challenge may be more preferable when the purpose is to confirm the diagnosis, instead of ruling out asthma (Coates et al., 2017).

Bronchial challenge testing should be performed by a technician or respiratory scientist well trained to treat acute bronchospasm. The patient must be able to understand and perform an acceptable and reproducible spirometry. In adults, it is not recommend to perform challenge test when the patient has pre-bronchodilator FEV_1 less than 60% predicted or 1.5 L or when the FEV_1 below 75% predicted for tests with a single high stimulus such as exercise (Coates et al., 2017). The low value of FEV_1 can lead to misinterpretation of the results. The tidal breath inhalation is the best while performing methacholine challenge, rather than total lung capacity. Tidal breath method has a higher sensitivity for a diagnosis of asthma (Cockcroft et al., 2020). While performing methacholine challenge test, initial spirometry maneuver is conducted in order to evaluate the baseline FEV_1. When a pre-spirometry test is completed, methacholine challenge test begins with nebulization, after which the FEV_1 measurement is recorded again. Subsequent increment doses of the methacholine are given to the patient according to a stepwise dosing protocol until a drop of 20% of the FEV_1 is obtained or the maximum dose step is reached (Kaplan et al., 2009). Direct airway challenges with methacholine are suggested as a diagnostic means, to increase or decrease the probability of asthma. However, evaluation of the performance of these tests is limited due to the absence of an independent gold standard method for objective confirmation of asthma (Coates et al., 2017). In addition, bronchial challenge tests are not appropriate to the young children or elderly patients as the tests require consistent spirometry test along with consistent inhalation of provoking agents. In this category of people, it is difficulty to perform spirometry or to do the exercise on the treadmill or riding a stationary bike.

3.1.4.2 Measurement of fractional exhaled nitric oxide

Airway inflammation is one of the central processes of asthma. Fractional exhaled nitric oxide (FENO) level test deals with measurement of the amount of nitric oxide (NO) exhaled while breathing. Measurement of the FENO has been proposed as a novel method to monitor different lung disease including asthma, chronic obstructive pulmonary disease and interstitial lung disease (Exhaled, 2005). After the description of the presence of endogenous nitric oxide in the exhaled air of animals and humans in 1991 (Gustafsson et al., 1991), several studies have reported elevated levels of the concentrations of fractional orally exhaled NO in patients with asthma as compared with non-asthmatic subjects. In asthma, the FENO has been suggested as a good marker of eosinophilic airway inflammation. The FENO values are higher in patients with eosinophilic airway inflammation as compared to those without eosinophilic asthma (Price et al., 2013).

There are two different methods of measuring the exhaled NO, specifically online and offline methods. With online methods, the NO analyzer continuously monitors the NO in the exhaled gas and displays in real time, the obtained NO profile with respect to time or exhaled volume, along with other exhalation parameters namely the airway flow rate and/or pressure. Online methods enable immediate evaluation of the required flow rate and pressure and thus, obtaining a suitable NO plateau. Defective exhalations are instantly identified and excluded (Exhaled, 2005). On the other hand, measurement of exhaled NO via offline methods, involves collection of the exhaled gas in a reservoir and then the concentrations of NO are analyzed (Exhaled, 2005). Advantages of offline methods include being independent from analyzer response time and facilitating to collect exhaled gas at the sites where the NO analyzer is not readily available such as in the clinics, remote laboratories, schools, and workplaces. However, with offline methods, collected exhaled gas can be contaminated with gases that are not originating from the lower airways. There is also a risk of error caused by the sample storage (Exhaled, 2005).

During measurement, the patient is instructed to sit comfortably. Then, the mouthpiece is inserted to the mouth of the patient, and she/he is requested to inhale deeply through the mouth for at least 2—3 seconds and then exhales immediately, because when the patient holds breath, it may affect the FENO measurements. The inhalation is done to total lung capacity (TLC), but inhalation near TLC is also acceptable in case the TLC is difficult to perform (Exhaled, 2005). The nose clip is not used in order to prevent the nasal NO to mix with the exhaled gas (Matsunaga et al., 2021). When the NO concentrations are high in the inhaled gas, an early NO peak appears in exhaled NO profile with respect to time. The time required to wash out this peak may be prolonged and it subsequentially increases the time required to reach a plateau, and thus, the duration for exhalation is extended (Exhaled, 2005).

While collecting expirate, measurement conditions should be kept constant. The concentrations of FENO depend on the expiratory flow rate, oral cavity pressure and lung volume during exhalations. The variation of the exhaled NO concentrations is inversely proportional to the variation in the flow rate. It means that as the flow rate decreases, the NO concentrations increase (Matsunaga et al., 2021). A flow rate of 50 mL/second must be maintained constant during measurement (Dweik et al., 2011) and it is acceptable and reproducible for both children and adults. But higher or lower flow rate can be considered depending on the situation. The acceptable range for expiratory flow rate is from 45 to 55 mL/s (Exhaled, 2005). The ATS/ERS guidelines proposed the exhalation pressures between 5 and 20 cm of H_2O to be applied during measurement. Pressures ranging between 5 and 15 cmH_2O were suggested by Japanese Respiratory Society (JRS) guidelines as they can close the soft palate by increasing the oral cavity pressure and then, the nasal NO derived in the upper airways is isolated from the NO produced in the lower airway (Matsunaga et al., 2021).

Measurement of exhaled NO is extensively applied in clinical settings for the assessment of airway inflammation (Korn et al., 2010). In this regard, Exhaled NO measurement can provide supplemental information to the conventional pulmonary function testing in a diagnosis of asthmatic patients with non-specific respiratory symptoms (Exhaled, 2005). It is also useful to predict response to inhaled corticosteroid (ICS) (Bjermer et al., 2014). The FeNO >50 parts per billion (ppb) has been reported to reflect a good short-term response to ICS. However, in those studies, long-term risks of exacerbations were not taken into consideration (Global Initiative for Asthma, 2021). Despite numerous advantages, measurement of the exhaled NO is not recommended as a useful tool for ruling in or ruling out a diagnosis of asthma. Because in asthma, the FENO is elevated in the one characterized by Type 2 airway inflammation and in non–asthma conditions such as eosinophilic bronchitis, atopy, allergic rhinitis, eczema, the value of FENO is also high. On contrarily, the FENO is low in some asthma phenotypes like neutrophilic asthma (Global Initiative for Asthma, 2021).

3.2 Capnography: a new approach for a diagnosis of asthma

There is a need for developing noninvasive tools for a diagnosis of asthma, especially in young children, elderly patients and patients suffering from chest pain, as they cannot easily perform objective diagnostic tests using spirometer. In addition, this category of patients cannot also easily tolerate bronchial challenge testing as it also involves consistently spirometry maneuver. Measurement of the exhaled NO is found useful for the assessment of airway inflammation, but it cannot be used as a diagnostic tool to rule in or rule out asthma. Therefore, capnography can be used as an alternative method to make a diagnosis of asthma and monitor the health condition of patient with airway obstruction. Capnography is different from both spirometry and peak flow rate measurement as the patient does not apply any extra effort during inspiration and expiration (Hisamuddin et al.,

2009). Capnography involves non-invasive measurement of the concentration of CO_2 in the respiratory breaths (Babik et al., 2012) and the CO_2 waveform produced is known as capnogram. Capnogram contains useful information for the diagnosis of the respiratory airway diseases including asthma and chronic obstructive pulmonary disease (COPD) (Mieloszyk et al., 2014). In addition, the analysis of the CO_2 waveform helps to monitor the ventilation status of the patient with respiratory distress, extubation outcomes, and the response to the treatment (Jaffe & Orr, 2010).

3.2.1 History of capnograph

Capnometry, discrete measurements of carbon dioxide concentration, was first developed for the purpose of monitoring the internal environment of submarines during the second world war (Egleston et al., 1997). In 1950s, capnometry entered clinical practice specifically in the experimental study carried out for measuring the expiratory CO_2 during anesthetic procedure (Hisamuddin et al., 2009). Later, in 1957, Smalhout began to use capnography while working in the Central Military Hospital of Utrecht, in Netherlands. In 1964, Smalhout recorded about 6000 manually annotated capnograms of different shapes (Smalhout & Kalenda, 1981). Thereafter, Smalhout and Kalenda published an atlas of strip-chart capnograms composed of more than 20 sections of annotated CO_2 waveforms versus time (Jaffe, 2017). Unlike capnometry, which displays only the numerical readouts, capnography displays the CO_2 waveform (known as capnogram) along with different parameters including end-tidal cardon dioxide (EtCO2), respiratory rate (RR) and the peripheral capillary oxygen saturation (SpO2) (Medtronic, 2022). Capnography is broadly used in different clinical situations such as monitoring ventilation status in patients with impaired respiratory function, verification of endotracheal tube placement, and the assessment of pathologic conditions such as bronchospasm and airway obstruction (Miller-Hance, 2019).

3.2.2 Basic principles of capnography

Capnography continuously monitors the concentration of CO_2 in the exhaled air. Capnography device known as capnograph is composed of different components and the predominant one is the CO_2 sensor. The CO_2 sensors that are mostly used includes chemical gas sensors and infrared (IR) gas sensors. Chemical gas sensors are built with sensitive layers, which are based on polymer or heteropolysiloxane. These sensors find applications in microelectronic based system due the miniaturized size in addition to the very low power consumption. However, their lifetime is short, and they require frequent calibration to maintain their stability (Azosensors, 2013). Infrared (IR) gas sensors especially non-dispersive infrared (NDIR) type are the most frequently used sensors due to their advantages such as the ability of uniquely detecting CO_2 without interference of other gases, resistance to high gas concentration exposure, and long lifetime.

The working principle of the infrared CO_2 sensors basically relies on the fact that CO_2 absorbs the infrared light at the wavelengths of 4.26 μm (Bhavani-Shankar & Philip, 2000). A basic NDIR sensor has an IR light source at one side, a gas chamber, an infrared filter, and an IR detector on the other side. As the respiratory gases pass through the sensor, the molecule of the exhaled CO_2 absorbs the IR light, and thus, decreasing the amount of IR light that reaches the infrared detector (Gallagher, 2018). The quantity of the light absorbed is proportional to the concentration of molecules available in the absorbing gas (Bhavani-Shankar & Philip, 2000). Therefore, the concentration of the CO_2 gas is calculated by considering the difference between the light emitted by the source and the light received by the detector (Gallagher, 2018). The measured CO_2 concentration is then displayed as the partial pressure of the end-tidal carbon dioxide expressed in mmHg, however, some units expressed it in percentage CO_2 (Bhavani-Shankar & Philip, 2000). Unlike capnometry, capnography displays the numerical values and the CO_2 waveform.

3.3 Types of capnographs

There are two types of capnographs that are currently used in the clinical practice namely mainstream capnograph and side stream capnograph (Fig. 3.5).

In the mainstream capnograph, the CO_2 sensor is placed closely to the subject between the endotracheal tube and the processing unit (Balogh et al., 2016). The location of the CO_2 sensor allows fast and appropriate measurement of the exhaled CO_2 that results in a quality signal (capnogram) (Jaffe, 2002). However, inappropriate connection of other components of the breathing circuit may cause the distortion of the signal. Mainstream capnographs have fast response time and the measurement results are obtained instantaneously. Although, these type of capnographs are only suitable for intubated patients as the mainstream sensors are relatively heavy and huge and frequent sterilization is required (Baba et al., 2020). In addition, these

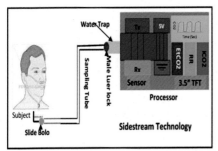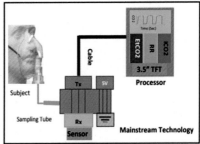

Figure 3.5 Types of capnographs: (A) sidestream method incorporating a sample tube, scavenge, water trap, water-permeable tube, CO_2 sensor, pump, processor, and monitoring unit; (B) the mainstream method incorporates a processor and monitoring unit (Singh et al., 2018).

sensors are subject to the internal heating above the core body temperature, which can cause facial burn of the patient (Baba et al., 2020). Recent mainstream sensors have more slight weight and disposable airway adapters are available. The temperature of the sensors is also reduced with an improved shielding and thus, facial burns are minimize (Kodali & Urman, 2014).

On the other hand, the side stream capnograph uses infrared sensor placed in the monitoring unit. The exhaled air from the subject is transported to the CO_2 sensor through the sampling plastic tube and mostly nasal cannula are used for this purpose (Maclennan & McCurry, 2020). The CO_2 samples are continuously monitored from the breathing circuit through the sampling tube and into the sample cell within the monitor at sample flow rates ranging from 50 to 250 mL/min (Jaffe, 2002). Stable monitoring of the CO_2 can be difficult due to the blockage of the sampling tube from moisture or secretions (Nagoshi et al., 2016). In the current designs of sidestream capnographs, a water trap is incorporated to the device and it is connected to the main unit to capture patient secretions and condensate that can damage the sensor (Jaffe, 2002) and thus, the shape of signal can be preserved.

Sidestream is the most commonly method for monitoring continuously the concentration of CO_2 that is exhaled in the breath (Kodali & Urman, 2014). However, their adequate use requires to properly maintain both their external physical setup and internal components. Side stream capnographs are suitable for intubated and non–intubated subjects and they are also convenient for both adults and children. However, collecting CO_2 samples using nasal cannula can result in irregular CO_2 waveform as compared to the facemask (e.g., cap-ONE) that is mostly used in mainstream devices

(Napolitano et al., 2017). Modified nasal cannulas are available and they can be used to enable exhaled gases to be sampled even during the time of administering supplemental oxygen (Kodali & Urman, 2014; Napolitano et al., 2017). Besides, sidestream capnographs have a simple connection and there is no difficulty in sterilization compared to the mainstream capnographs (SIngh & MalarvIll, 2018). The drawback of the sidestream capnograph, is that the movement of gas through the sampling tube causes the delay in response time, which subsequently increases the time in the detection of the CO_2 concentration (Balogh et al., 2016). There is also a risk of dispersion inside of the sampling tube resulting from the effect of velocity profile and diffusion. Moreover, the sample flow rate may be affected by different factors such as the length of the sampling tube and airway pressure (Jaffe, 2002).

References

Adeniyi, B., & Erhabor, G. (2011). The peak flow meter and its use in clinical practice. *African Journal of Respiratory Medicine, 6*(2), 5–7.

Azosensors. (2013). *Carbon dioxide sensor.* Retrieved August 04 from https://www.azosensors.com/article.aspx?ArticleID=234

Baba, Y., Takatori, F., Inoue, M., & Matsubara, I. (2020). A novel mainstream capnometer system for non-invasive positive pressure ventilation. In *2020 42nd Annual International Conference of the IEEE Engineering in Medicine & Biology Society (EMBC).*

Babik, B., Csorba, Z., Czövek, D., Mayr, P. N., Bogáts, G., & Peták, F. (2012). Effects of respiratory mechanics on the capnogram phases: Importance of dynamic compliance of the respiratory system. *Critical Care, 16*(5), R177.

Balogh, A., Peták, F., Fodor, G., Tolnai, J., Csorba, Z., & Babik, B. (2016). Capnogram slope and ventilation dead space parameters: Comparison of mainstream and sidestream techniques. *BJA: British Journal of Anaesthesia, 117*(1), 109–117.

Barreiro, T., & Perillo, I. (2004). An approach to interpreting spirometry. *American Family Physician, 69*(5), 1107–1114.

Bhavani-Shankar, K., & Philip, J. H. (2000). Defining segments and phases of a time capnogram. *Anesthesia and Analgesia, 91*(4), 973–977.

Bjermer, L., Alving, K., Diamant, Z., Magnussen, H., Pavord, I., Piacentini, G., Price, D., Roche, N., Sastre, J., & Thomas, M. (2014). Current evidence and future research needs for FeNO measurement in respiratory diseases. *Respiratory Medicine, 108*(6), 830–841.

Castro, M., & Kraft, M. (2008). *Clinical asthma e-book.* Elsevier Health Sciences.

Chung, K. S., Jung, J. Y., Park, M. S., Kim, Y. S., Kim, S. K., Chang, J., & Song, J. H. (2016). Cut-off value of FEV1/FEV6 as a surrogate for FEV1/FVC for detecting airway obstruction in a Korean population. *International Journal of Chronic Obstructive Pulmonary Disease, 11*, 1957.

Ciprandi, G., Gallo, F., & Cirillo, I. (2018). FEF25-75 and asthma in clinical practice. *Iranian Journal of Allergy, Asthma and Immunology,* 295–297.

Coates, A. L., Wanger, J., Cockcroft, D. W., Culver, B. H., Carlsen, K.-H., Diamant, Z., Gauvreau, G., Hall, G. L., Hallstrand, T. S., & Horvath, I. (2017). ERS technical standard on bronchial challenge testing: General considerations and performance of methacholine challenge tests. *European Respiratory Journal, 49*(5).

Cockcroft, D. W., Davis, B. E., & Blais, C. M. (2020). Comparison of methacholine and mannitol challenges: Importance of method of methacholine inhalation. *Allergy, Asthma & Clinical Immunology, 16*(1), 1–12.

Dempsey, T. M., & Scanlon, P. D. (2018). Pulmonary function tests for the generalist: A brief review. *Mayo Clinic Proceedings, 93*(6), 763–771.

DeVrieze, B. W., Modi, P., & Giwa, A. O. (2017). *Peak flow rate measurement.*

Dweik, R. A., Boggs, P. B., Erzurum, S. C., Irvin, C. G., Leigh, M. W., Lundberg, J. O., Olin, A.-C., Plummer, A. L., Taylor, D. R., & American Thoracic Society Committee on Interpretation of Exhaled Nitric Oxide Levels (Fe_{NO}) for Clinical Applications. (2011). An official ATS clinical practice guideline: Interpretation of exhaled nitric oxide levels (FENO) for clinical applications. *American Journal of Respiratory and Critical Care Medicine, 184*(5), 602–615.

Education, N. A., & Program, P. (2007). Expert panel report 3 (EPR-3): Guidelines for the diagnosis and management of asthma-summary report 2007. *The Journal of Allergy and Clinical Immunology, 120*(5 Suppl. l), S94–S138.

Egleston, C., Aslam, H. B., & Lambert, M. (1997). Capnography for monitoring non-intubated spontaneously breathing patients in an emergency room setting. *Emergency Medicine Journal, 14*(4), 222–224.

Exhaled, N. (2005). ATS/ERS recommendations for standardized procedures for the online and offline measurement of exhaled lower respiratory nitric oxide and nasal nitric oxide. *American Journal of Respiratory and Critical Care Medicine, 171*(8), 912–930.

Gallagher, J. J. (2018). Capnography monitoring during procedural sedation and analgesia. *AACN Advanced Critical Care, 29*(4), 405–414.

Global Initiative for Asthma. (2021). *Global strategy for asthma management and prevention.* www.ginasthma.org

Graham, B. L., Steenbruggen, I., Miller, M. R., Barjaktarevic, I. Z., Cooper, B. G., Hall, G. L., Hallstrand, T. S., Kaminsky, D. A., McCarthy, K., & McCormack, M. C. (2019). Standardization of spirometry 2019 update. An official American thoracic society and European respiratory society technical statement. *American Journal of Respiratory and Critical Care Medicine, 200*(8), e70–e88.

Gustafsson, L. E., Leone, A., Persson, M., Wiklund, N., & Moncada, S. (1991). Endogenous nitric oxide is present in the exhaled air of rabbits, Guinea pigs and humans. *Biochemical and Biophysical Research Communications, 181*(2), 852–857.

Guyton, A. C., & Hall, J. E. (1986). *Textbook of medical physiology, 548.* Philadelphia: Saunders.

Haynes, J. M. (2018). Basic spirometry testing and interpretation for the primary care provider. *Canadian Journal of Respiratory Therapy, 54*(4).

Hazari, M., & Costa, D. (2010). Pulmonary mechanical function and gas exchange. *Comprehensive Toxicology: Respiratory Toxicology, 151*–169.

Hisamuddin, N. N., Rashidi, A., Chew, K., Kamaruddin, J., Idzwan, Z., & Teo, A. (2009). Correlations between capnographic waveforms and peak flow meter measurement in emergency department management of asthma. *International Journal of Emergency Medicine, 2*(2), 83–89.

Hyatt, R. E., Scanlon, P. D., & Nakamura, M. (2014). *Interpretation of pulmonary function tests.* Lippincott Williams & Wilkins.

Jaffe, M. B. (2002). Mainstream or sidestream capnography. *Environment, 4*(5).

Jaffe, M. B. (2017). Using the features of the time and volumetric capnogram for classification and prediction. *Journal of Clinical Monitoring and Computing, 31*(1), 19–41.

Jaffe, M. B., & Orr, J. (2010). Continuous monitoring of respiratory flow and Co2. *IEEE Engineering in Medicine and Biology Magazine, 29*(2), 44–52.

Jat, K. R. (2013). Spirometry in children. *Primary Care Respiratory Journal, 22*(2), 221–229.

Jerzyńska, J., Janas, A., Galica, K., Stelmach, W., Woicka-Kolejwa, K., & Stelmach, I. (2015). Total specific airway resistance vs spirometry in asthma evaluation in children in a large real-life population. *Annals of Allergy, Asthma & Immunology, 115*(4), 272–276.

Johns, D. P., Walters, J. A., & Walters, E. H. (2014). Diagnosis and early detection of COPD using spirometry. *Journal of Thoracic Disease, 6*(11), 1557.

Kaplan, A. G., Balter, M. S., Bell, A. D., Kim, H., & McIvor, R. A. (2009). Diagnosis of asthma in adults. *CMAJ, 181*(10), E210–E220.

Kavanagh, J., Jackson, D. J., & Kent, B. D. (2019). Over-and under-diagnosis in asthma. *Breathe, 15*(1), e20–e27.

Khandpur, R. S. (2005). *Biomedical instrumentation.*

Kodali, B. S., & Urman, R. D. (2014). Capnography during cardiopulmonary resuscitation: Current evidence and future directions. *Journal of Emergencies, Trauma, and Shock, 7*(4), 332.

Korn, S., Telke, I., Kornmann, O., & Buhl, R. (2010). Measurement of exhaled nitric oxide: Comparison of different analysers. *Respirology, 15*(8), 1203–1208.

Levy, M. L., Quanjer, P. H., Rachel, B., Cooper, B. G., Holmes, S., & Small, I. R. (2009). Diagnostic spirometry in primary care: Proposed standards for general practice compliant with American thoracic society and European respiratory society recommendations. *Primary Care Respiratory Journal, 18*(3), 130–147.

Maclennan, T., & McCurry, R. (2020). Capnography-what is it all about? *Veterinary Nursing Journal, 35*(8), 231–234.

Matsunaga, K., Kuwahira, I., Hanaoka, M., Saito, J., Tsuburai, T., Fukunaga, K., Matsumoto, H., Sugiura, H., Ichinose, M., & Japanese Respiratory Society Assembly on Pulmonary Physiology. (2021). An official JRS statement: The principles of fractional exhaled nitric oxide (FeNO) measurement and interpretation of the results in clinical practice. *Respiratory Investigation, 59*(1), 34–52.

Medtronic. (2022). *Capnostream™ 20p bedside monitor with Apnea-Sat alert algorithm.* Retrieved 04 April from https://www.medtronic.com/covidien/en-us/products/capnography/capnostream-20p-bedside-patient-monitor.html

Mieloszyk, R. J., Verghese, G. C., Deitch, K., Cooney, B., Khalid, A., Mirre-González, M. A., Heldt, T., & Krauss, B. S. (2014). Automated quantitative analysis of capnogram shape for COPD–normal and COPD–CHF classification. *IEEE Transactions on Biomedical Engineering, 61*(12), 2882–2890.

Miller-Hance, W. C. (2019). Anesthesia for noncardiac surgery in children with congenital heart disease. In *A practice of anesthesia for infants and children.* Elsevier, 534.e539–559.e539.

Nagoshi, M., Morzov, R., Hotz, J., Belson, P., Matar, M., Ross, P., & Wetzel, R. (2016). Mainstream capnography system for nonintubated children in the postanesthesia care unit: Performance with changing flow rates, and a comparison to side stream capnography. *Pediatric Anesthesia, 26*(12), 1179–1187.

Napolitano, N., Nishisaki, A., Buffman, H. S., Leffelman, J., Maltese, M. R., & Nadkarni, V. M. (2017). Redesign of an open-system oxygen face mask with mainstream capnometer for children. *Respiratory Care, 62*(1), 70–77.

National Heart, L., & Institute, B. (2007). *National asthma education and prevention program. Expert panel report 3: Guidelines for the diagnosis and management of asthma: Full report 2007.* http://www.nhlbi.nih.gov/guidelines/asthma/asthgdln.pdf

Nazir, Z., Razaq, S., Mir, S., Anwar, M., Al Mawlawi, G., Sajad, M., Shehab, A., & Taylor, R. (2005). Revisiting the accuracy of peak flow meters: A double-blind study using formal methods of agreement. *Respiratory Medicine, 99*(5), 592–595.

Price, D., Ryan, D., Burden, A., Von Ziegenweidt, J., Gould, S., Freeman, D., Gruffydd-Jones, K., Copland, A., Godley, C., & Chisholm, A. (2013). Using fractional exhaled nitric oxide (FeNO) to diagnose steroid-responsive disease and guide asthma management in routine care. *Clinical and Translational Allergy, 3*(1), 1–10.

Singh, O. P., Howe, T. A., & Malarvili, M. (2018). Real-time human respiration carbon dioxide measurement device for cardiorespiratory assessment. *Journal of Breath Research, 12*(2), 026003.

SIngh, O. P., & MalarvIII, M. (2018). Review of infrared carbon-dioxide sensors and capnogram features for developing asthma-monitoring device. *Journal of Clinical & Diagnostic Research, 12*(10).

Smalhout, B., & Kalenda, Z. (1981). *An atlas of capnography*. Utrecht: Institute of Anaesthesiology, University Hospital.

Sokol, K. C., Sharma, G., Lin, Y.-L., & Goldblum, R. M. (2015). Choosing wisely: Adherence by physicians to recommended use of spirometry in the diagnosis and management of adult asthma. *The American Journal of Medicine, 128*(5), 502–508.

South Bay Allergy and Asthma. (2022). *Peak flow meter*. Retrieved 13 March from https://www.southbayallergy.com/peak-flow-meter.html

Thaslima, N. S., & Gayatri, D. R. (2020). A comparative study of peak expiratory flow rate in acute and chronic periodontitis. *Journal of Evolution of Medical and Dental Sciences, 9*(44), 3294–3300.

Thompson, J. E., & Jaffe, M. B. (2005). Capnographic waveforms in the mechanically ventilated patient. *Respiratory Care, 50*(1), 100–109.

Wanger, J. (2012). *Pulmonary function testing. A practical approach* (3rd ed., p. 21). Burlington, MA: Jones & Bartlett Learning.

van den Wijngaart, L. S., Roukema, J., & Merkus, P. J. (2015). The value of spirometry and exercise challenge test to diagnose and monitor children with asthma. *Respirology Case Reports, 3*(1), 25–28.

CHAPTER FOUR

Capnography: measurement of expired carbon dioxide

The cellular metabolism produces CO_2 in the body and circulatory system conveys to the lungs, expelled by the lungs, and forwarded by the breathing system. Therefore, it is believed that the expired CO_2 alternation may provide the information about the airway passage blockages, circulation, and metabolism, breathing respiration, or breathing system. There are several reasons that cause increase in CO_2, which are administration of blood, shivering, convulsions, and many others. Capnography is noninvasive method, measures human expired CO_2, and has various applications to know the status of circulatory system. The CO_2 waveform produced is known as capnogram. Graphical representation of exhaled CO_2 can be expressed either as a temporal series while using a time-based capnography or as a function of exhaled volume via volumetric capnography. This book focuses on time-based capnography and the CO_2 waveform is referred as CO_2 signal. The terms CO_2 waveform, CO_2 signal, and capnogram are used interchangeably throughout the book. The capnogram and associated features are also useful to know the metabolic CO_2 status from the exhaled gas continuously. Capnography also displays the spontaneous respiratory rate in nonintubated patients breathing that allows the detection of airway obstruction and apnea. Carbon dioxide is part of the most essential molecules found in human body, and it is inherently linked with different processes including metabolism, circulation, and ventilation. Therefore, this chapter will provide the overview about CO_2 removal, gas exchange in the lungs, ventilation/perfusion inequality, measurement of CO_2 in the blood, PetCO2 monitoring, and graphical representation of the exhaled volume. Lastly, more in-depth discussion about normal capnogram interpretation and capnogram abnormalities is provided.

4.1 Carbon dioxide production and removal from human body

The earlier discussion in Chapter 1 denoted that CO_2 is produced in all cells of the body as a result of cellular metabolism (Chambers et al., 2019), which is the process that converts the glucose and nutrients to energy,

Systems and Signal Processing of Capnography as a Diagnostic Tool for Asthma Assessment
ISBN: 978-0-323-85747-5
https://doi.org/10.1016/B978-0-323-85747-5.00005-X

needed for the cells to live, using oxygen (O_2) and producing CO_2 and water as a by-product. Ingested carbohydrates are converted to glucose. Breakdown of glucose is a chemical reaction and the number of O_2 molecules consumed is equal to the number of CO_2 produced as expressed in the chemical Eq. (4.1) (Gravenstein et al., 2011):

$$C_6H_{12}O_6 + 6O_2 \rightarrow 6CO_2 + 6H_2O + energy \qquad (4.1)$$

Respiratory and cardiovascular systems work interactively to ensure that the cells of the body are satisfactory oxygenated, and proper amount of CO_2 is removed (Gravenstein et al., 2011). At standard temperature and pressure, the CO_2 produced varies between 100 and 300 mL/min in adults and in mechanically ventilated patients under heavy sedation or general anesthesia, this value is decreased by 15%–20% (Kremeier et al., 2020). In neonates, CO_2 production is extremely lower (about 15 mL/min) compared to adults (Schmalisch, 2016). During intense exercise, CO_2 production can increase up to 4000 mL/min (Chambers et al., 2019). Because the muscles moving in the body need lots of energy. The metabolic rate of the muscles specifically, and the whole body in general, increases several fold. Thus, the breathing increases, to get in more O_2 and deliver it to the cells for the increased metabolism; and at the same time, to rapidly excrete out all the added CO_2 that is being produced (Gravenstein et al., 2011).

In addition, when consuming a large portion of meal, for example, the metabolic rate increases to meet the needs of digesting and storing the nutrients from that large meal (Calcagno et al., 2019). The higher metabolic rate produces more CO_2, but since it is produced at a relatively slow rate, the excretion of CO_2 increases only via the kidneys into the urine. The breathing rate is usual after a normal meal to excrete CO_2. In a cold environment, metabolic rate is also increased through tensing muscles and shivering so that the body heat can be maintained (Gravenstein et al., 2011). Throughout the day, cells in an organism are using O_2 to burn glucose producing energy and CO_2 as a waste product. Thus, continuous removal of CO_2 from the body is one of the main requirements of the human being (Kremeier et al., 2020). If too much CO_2 is left within the body, it will increase the acidity of internal environment, which will be a danger to human life (Patel & Sharma, 2021). Chronic respiratory acidosis may lead to several abnormal health conditions including pulmonary hypertension, memory loss, hypercapnia, and heart failure (Patel & Sharma, 2021). Respiratory acidosis has also been identified in asthmatic patients with severe airway obstruction (Raimondi et al., 2013).

Luckily, human body is highly adept at ensuring that most of the CO_2 produced is finally excreted from the body, maintaining a fine balance to the internal environment. This is done in two main ways. First, much of the CO_2 is converted into bicarbonates, which is then excreted via the kidneys in the urine (Neligan & O'Donoghue, 2010). This is a high-capacity system, meaning that it is able to excrete large quantities of CO_2 in various forms. But the excretion of CO_2 via the kidneys is not a rapid process. It takes hours to days. In short, the kidneys are a high capacity but slower mode of CO_2 excretion. Second, great amount of the CO_2 produced within the cells is removed through the lungs with each exhaled breath. Unlike the kidneys, the excretion of CO_2 through the lungs is a lower capacity system. It is limited by how often and rapid is the breathing rate. Despite being limited by the limits of breathing, this system is very important because it works very quickly, within minutes. This low capacity, rapid responding system allows the minute-to-minute control of CO_2 levels within the body. Elimination of CO_2 via the lungs can be increased or decreased due to different factors such as hyperventilation, hypoventilation, and status of cardiac output. Acute hyperventilation resulting from anxiety or discomfort causes a transient increase in pulmonary CO_2 elimination and reduction in $PaCO_2$. Conversely, hypoventilation results in a transient decrease in CO_2 elimination, but reverses once the alveolar CO_2 concentration rises to a new steady state. Transient decrease in the CO_2 elimination may also result from a decreased cardiac output (Gravenstein et al., 2011).

4.2 Overview of pulmonary gas exchange

Carbon dioxide is the predominant product of cellular respiration (Gravenstein et al., 2011). After gas exchange, the CO_2 diffused in the alveoli, is conveyed outside the body through respiratory upper airways. Understanding the flow of gases between alveoli and pulmonary capillaries and inequalities in alveolar ventilation and perfusion will facilitate the analysis and interpretation of different phases of a CO_2 waveform in normal and abnormal respiratory conditions. Because each phase of a CO_2 waveform reflects the usual process of CO_2 excretion. Gas exchange results from differences in partial pressures of oxygen and carbon dioxide in which gas moves from areas of higher pressure to lower pressure (Aung et al., 2019). Fig. 4.1 indicates how the gases diffuse between alveoli and bloodstream in the alveolar capillary membrane. Exchange of gases between alveolar air and pulmonary capillary blood is done by passive diffusion. Pulmonary arterial blood

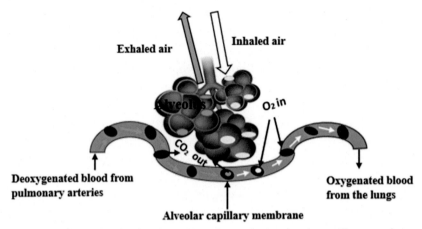

Figure 4.1 Illustration of pulmonary gas exchange in the alveolar capillary membrane.

that comes from the tissues to the capillary in the alveolar wall, contains elevated amount of CO_2 and less amount of O_2. Then the CO_2 diffuses to the alveoli, whereas the diffusion of O_2 is in the opposite direction; it means from alveoli to the alveolar capillary membrane (Aung et al., 2019). Thereafter, pulmonary veins conveys the blood rich in O_2 but poor in CO_2 back to the heart firstly in the left atrium (Wagner, 2015). In normal lungs that are effectively perfused, pulmonary capillary and alveolar CO_2 comes into equilibrium in <0.5 seconds, but the time is extended when there is a certain lung pathology (Ward & Yealy, 1998a). The movement of gas molecules across alveolar capillary membrane is determined by the partial pressure difference between the alveoli and the pulmonary capillary blood. Carbon dioxide diffusion is faster compared to that of oxygen. For a given pressure difference, CO_2 diffuses approximately 20 times as quickly as O_2. The reason is that the solubility of CO_2 in the water is 30 times higher than the one of O_2 (Qureshi & Mustafa, 2018). The process of gas exchange is influenced by many factors including respiratory membrane thickness, alveolar–capillary contact surface area, gas barrier and ventilation to perfusion mismatch in the lung regions (Aung et al., 2019). When the thickness of the respiratory membrane increases, the rate of diffusion through the membrane decreases. For example, due to pulmonary edema, respiratory membrane thickness rises above the normal value for more than two to three times, which can result in a great interruption in normal gas exchange. In addition, thickened alveolar capillary membrane will extent the gas exchange equilibrium and may lead to hypoxemia (Qureshi & Mustafa, 2018).

Pulmonary gas exchange is a vital process through which oxygen is supplied to the body cells and carbon dioxide is transported outside the body. Gas exchange takes place in the alveoli of the lungs via diffusion. During inhalation, oxygen passes in the respiratory airways and continues to the lungs. After reaching alveoli, O_2 diffuses to the pulmonary blood capillary and then exits the lungs to the heart though pulmonary veins. On the other hand, CO_2 produced during cellular metabolism, leaves from the tissue traveling in the blood through vena cava to the heart that pumps it to the lungs via pulmonary arteries. Then CO_2 diffuses to the alveoli, and subsequently leaves the body through respiratory bronchioles and the upper respiratory tract (Albert et al., 2008). During respiration, not all the amount of the inspired air enters alveoli. There is a portion of ventilation known as anatomical dead space that remains in the conducting zones and it does not participate to the exchange of gas. The same thing happens to the alveolar dead space, which is the fraction of ventilation found in the alveoli that are ventilated with air but not perfused with pulmonary arterial blood (Robertson, 2015). Anatomical and alveolar dead space together give physiological dead space, which is considered as wasted ventilation (Albert et al., 2008).

During normal breathing, the average partial pressure of alveolar carbon dioxide ($PaCO_2$) is 40 mmHg. Hypoventilation relates to the decrease in alveolar ventilation, and it results in increased levels of the carbon dioxide production known as hypercapnia (Wolfson & Shaffer, 2004). Contrarily, hyperventilation is a condition in which alveolar ventilation increases and it causes reduction in the carbon dioxide levels at both side of alveolar capillary membrane (Kremeier et al., 2020). Hyperventilation leads to the depletion of CO_2 production in body referred to as hypocapnia (Tsakiris et al., 2021). Hypocapnia is a significant feature in the pathophysiology of asthma (van't Hul et al., 2018). Because it can cause detrimental effect such as rising central and peripheral airway resistance in the asthmatic patients (Van den Elshout et al., 1991). In a study by conducted (Deenstra et al., 2021) on 1006 patients with stable asthma, 227 patients were identified with chronic hyperventilation and the prevalence was higher (79%) in females of younger age. In this subgroup of patients, hyperventilation was assumed to be associated with the effect of over breathing. However, some patients with excess breathing did not experience hypocapnia. Therefore, validity of this assumption is still a point of concern (Deenstra et al., 2021).

4.3 Ventilation/perfusion inequalities

Pulmonary gas exchange involves three different processes that interact simultaneously. These are: ventilation, perfusion and diffusion (Kremeier et al., 2020). Diffusion facilitates proper oxygenation and appropriate elimination of the CO_2 from pulmonary capillary blood. The diffusion capacity is significantly reduced in disorders that affect pulmonary vessels and in chronic obstructive pulmonary diseases (Qureshi & Mustafa, 2018). Alveolar ventilation allows successful clearance of CO_2 in the whole-body (Ward & Yealy, 1998a). Imbalance of any of these processes interrupts homeostasis and causes hypercapnia or hypocapnia. Hypercapnia is mostly experienced by patients that are mechanically ventilated (Kremeier et al., 2020). The efficiency of gas exchange in various areas of the lung is evaluated based on the concept of ventilation/perfusion (V/Q) relationship (Kremeier et al., 2020; Pizano et al., 2019). The process by which the air flows into and out of the alveoli is known as ventilation, whereas perfusion implies the flow of blood in pulmonary capillary. Effective gas exchange requires alveoli ventilation and perfusion to be well matched (Powers & Dhamoon, 2019). When the V/Q ratio is about one, it means that both ventilation and perfusion matches. Unfortunately even in normal lung, ventilation and perfusion are not homogeneous in all 300 million alveoli (Wagner, 2015). When an individual is at rest, ventilation in the upper parts of the lungs is slightly higher compared to the perfusion; during exercise, the flow of blood toward the upper regions of the lungs rises, which decreases physiological dead space and thus more adequate gas exchange is achieved (Qureshi & Mustafa, 2018).

Any inequality in the V/Q ratio can possibly cause the variation in partial pressure of O_2 (PaO$_2$) and partial pressure of CO_2 (PaCO$_2$), which leads to the gas exchange impairment. Variations in the V/Q ratio and their implications to the PaO$_2$ PaCO$_2$ are summarized in Table 4.1. Reduction in the V/Q ratio in given lung region, is an indication that the region is poorly ventilated, but perfusion is normal. This condition occurs frequently due to the airway obstruction. As a result, PaCO$_2$ can be slightly increased but PaO$_2$ will be decreased. Contrarily, the rise of the V/Q in a lung region, usually results from vascular obstruction and it is associated with falling of the PaCO$_2$ and rising of the PaO$_2$ (Wagner, 2015).

Absence of alveolar ventilation in some regions of the lungs leads to the V/Q that is close to zero. In other areas, the blood flow is inefficient as a result of capillaries that are damaged and the alveolar ventilation is wasted

Table 4.1 The V/Q ratios and their implications to the partial pressure of gases (O_2 and CO_2) in the alveolar air (Qureshi & Mustafa, 2018).

Status	Ventilation (V)	Perfusion(Q)	V/Q ratio	Meaning
	Normal	Normal	Normal	Both ventilation and perfusion are normal for the same alveolus; alveolar air has: PaO_2 of 104 mmHg and $PaCO_2$ of 40 mmHg.
	Zero	Normal	Zero	Perfusion but no ventilation for the alveolus. Alveolar air equilibrates with mixed venous blood gases. PaO_2 of 40 mmHg and $PaCO_2$ of 45 mmHg.
	Normal	Zero	Infinity	Ventilation but no perfusion. No O_2 is lost, or no CO_2 is gained in the alveolar air. PaO_2 of 149 mmHg and $PaCO_2$ of 0 mmHg.

and thus the V/Q increases toward infinity (Qureshi & Mustafa, 2018). Hypoxaemia, and arterial hypercapnia are usually manifested as initial effects of the V/Q inequality. Basically all lung diseases lead to a great V/Q inequality, however the physiological and structural mechanisms are not common in all diseases (Wagner, 2015). Asthmatic patients may experience hypoxaemia due to ventilation/perfusion mismatch in a great number of lung regions that are poorly ventilated and obstructed (Taytard et al., 2020).

4.4 Measurement of carbon dioxide

Carbon dioxide is produced in all cells of the body, and it is excreted by the lungs and then to other parts of the respiratory system. Therefore, any problems with the lungs ability to remove CO_2 should be seen as changes in CO_2 levels in the blood. Though, the body compensates with any problems in the lungs by increasing breathing rates and increasing excretion via the kidneys. These can also be monitored from blood samples. CO_2 levels in the blood can be measured in arterial blood samples, and it is usually expressed as the partial pressure of CO_2 in mmHg or kPa. This is part of the commonly used arterial blood gases (ABG) analysis. ABG is currently used as a standard method to measure levels of CO_2 in the blood. However, this method has several limitations such as taking blood samples repeatedly when measurement of CO_2 levels is required. In addition, there is still no feasible method to continuously monitor the amount of CO_2 in blood by direct measurement. Monitoring exhaled CO_2 concentrations from the respiratory breath is a viable means to estimate arterial blood CO_2. However, this approach is more appropriate in normal lungs. The next sections discuss about measuring CO_2 levels in the blood via ABG analysis and its limitations, partial pressures of carbon dioxide (PCO_2) and $PetCO_2$ measurement, and measurement of exhaled CO_2 volume.

4.4.1 Measurement of CO_2 levels in the blood

Arterial blood gas (ABG) test is the most accurate and main method of CO_2 measurement in the blood. CO_2 levels and ABG results are vital measurements that give doctors a clearer picture on how oxygenation and ventilation processes are performing at any given time; or in short, how O_2 and CO_2 is being moved into and out of the body. This is a key function of life, and therefore a vital function to continuously monitor. While performing arterial blood gas (ABG) test, blood samples are taken from the artery and then, both partial pressure of oxygen (PaO_2) and carbon dioxide ($PaCO_2$) of the patient can be assessed. Analysis of the $PaCO_2$ results can provide information related to the patient's ventilation status with either chronic or acute respiratory failure. The $PaCO_2$ level is affected by hyperventilation (rapid or deep breathing), hypoventilation (slow or shallow breathing), and acid–base status (Castro et al., 2021).

Although oxygenation and ventilation can be assessed non–invasively using pulse oximetry and end–tidal carbon dioxide monitoring, respectively, ABG analysis is the gold standard test for measuring $PaCO_2$ (Castro et al., 2021; Dicembrino et al., 2021). However, this test has several limitations. In fact, the specimen used in this test are blood samples collected via arterial puncture (Castro et al., 2021), which can be the source of pain and anxiety especially for children. Other complications can also be manifested with the possibility of infection, injury of tissue, vessels, and nerves as well (Dicembrino et al., 2021). The ABG is not an easily available test because it requires arterial blood sampling each time a reading is needed, which is difficult to perform. The test is expensive and it requires to be performed in the laboratory by a trained personnel (Dicembrino et al., 2021). Besides, it only shows the CO_2 levels at that point in time, when the blood sample was taken (Kremeier et al., 2020). However, the CO_2 levels can rapidly increase and decrease. Thus, using the ABG as the only mode of measuring CO_2 levels is limited, since taking multiple ABG blood samples throughout the day is not viable in most situations. A CO_2 measurement that does not require taking blood samples would be better. A test that could measure CO_2 continuously would be fantastic. If the measurement of CO_2 levels is continuous without having to take blood samples from the patient repeatedly, it would be ideal. Unfortunately, as of today, there is still no viable way to continuously monitor the amount of CO_2 in blood by direct measurement. However, expired CO_2 concentrations can be measured easily from the exhaled breath.

4.4.2 PaCO$_2$ and PetCO$_2$ measurement

Analysis of the variation in the CO_2 concentration during inhalation and exhalation provides significant clinical information for monitoring ventilation status especially in patients with respiratory abnormalities. Currently, measurement of exhaled CO_2 in the respiratory breaths is regarded as standard practice in different clinical situations. Different researchers have quantitatively analyzed the modifications in the respiratory waveform during expiration and inspiration. In those studies, the expiratory portion is the one that is mostly reported to have valuable information reflecting the flow of respiratory gases particularly in the obstructive airways, which is the case for asthmatic patients. The motive for measuring partial pressure of CO_2 (PCO_2) in exhaled gas is based on the assumption that the end-tidal PCO_2 ($PetCO_2$) should mimic the arterial PCO_2 ($PaCO_2$) quite closely. However, these parameters are affected by different variables. Understanding the interrelationship between $PaCO_2$ and $PetCO_2$ will ease the understanding of the $PetCO_2$ measurement, interpretation of graphical representation of the exhaled CO_2, and differentiating normal from abnormal respiratory patterns, which will be discussed in the next sections of this chapter.

Arterial PCO_2 ($PaCO_2$) means the partial pressure of CO_2 in arterial blood and it has become a measure to evaluate the efficiency of ventilation (Ward & Yealy, 1998a). The $PaCO_2$ between 35 and 45 mmHg is accepted as a clinically normal range (Wagner, 2015). The variation in the arterial CO_2 tensions can be caused by respiratory diseases such asthma, and pulmonary edema. Besides, other clinical conditions including acute lung injury, excessive or inadequate mechanical ventilation, high-frequency ventilation, cardiopulmonary bypass, extracorporeal membrane oxygenation and narcotic analgesic administration can also alter the $PaCO_2$ measurement (Gravenstein et al., 2011). Fortunately, arterial PCO_2 can be normalized even by small compensatory rise in ventilation (Wagner, 2015). On the other hand, $PetCO_2$ or simply $EtCO_2$ means the end tidal partial pressure of CO_2 and it is the highest value of exhaled CO_2 in a respiratory breath. In a capnogram, $PetCO_2$ value is marked at the end of expiratory phase before the inspiratory phase begins. $PetCO_2$ is expressed in millimeters of mercury (mmHg) or parts per million (ppm) or in percentages (Malik et al., 2016; Singh et al., 2018). Assumption was made that $PetCO_2$ measurement can reflect arterial PCO_2. Clinicians soon realized that $PetCO_2$ readings closely mimicked arterial CO_2 concentrations only in normal lungs.

It is frequently assumed that in normal lungs, the $PetCO_2$ readings are very close to the blood CO_2 readings usually displaying a smaller reading by just a few mmHg (≤ 5 mmHg) (Kremeier et al., 2020), assuming that alveolar dead space is 12%–15% (Gravenstein et al., 2011). In mechanically ventilated, critically ill patients the $PaCO_2$–$PetCO_2$ difference is higher and it can sometimes be considered while monitoring this category of patients (Gravenstein et al., 2011). Hypoperfusion of the lung, a condition in which alveolar ventilation is higher than perfusion in many regions of the lungs causes impairment in the CO_2 transfer from pulmonary blood arteries to the lungs. Consequently, the measured $PetCO_2$ will be less as compared to the patient's true $PaCO_2$ (Rhoades & Thomas, 2002). Caution must be taken while approximating arterial blood gas CO_2 ($PaCO_2$) based on the $PetCO_2$ measurement (Surrey et al., 2018). Because, $PaCO_2$ is generally greater than $PetCO_2$ due to the effects of shunt and physiological dead space (Kremeier et al., 2020; Surrey et al., 2018).

The end-tidal PCO_2 is frequently measured in various clinical situation including the assessment of cardiopulmonary resuscitation (CPR) outcomes and return of spontaneous circulation status (Ruiz et al., 2016; Selby et al., 2018; Stine et al., 2019), monitoring respiratory status in non–intubated patients undergoing procedural sedation (Mehta et al., 2017), management of childhood seizures and diabetic ketoacidosis (Langhan et al., 2008). Moreover, Et CO_2 measurement is clinically useful to monitor airways status in patients with obstructive respiratory diseases including asthma and COPD (Aminiahidashti et al., 2018). Different researchers examined the usefulness of the $EtCO_2$ measurement for the disease severity estimation, patient outcomes prediction (Selby et al., 2018), evaluation of the effectiveness of the therapy, and prediction of hospitalization in asthmatic patients (Kunkov et al., 2005). For children with acute asthma aged between 3 and 17 years old, the $EtCO_2$ values failed to classify asthma severity levels (Guthrie et al., 2007). Reduction in the $Et\ CO_2$ value below the normal range can result from the decrease in the cardiac output and pulmonary blood flow (Gravenstein et al., 2011). The decrease in the $EtCO_2$ observed after a mechanical breath may reflect inappropriate distribution of the intrapulmonary gas (Carlon et al., 1988).

$EtCO_2$ measurement is valuable in view of monitoring and diagnosis of respiratory diseases. However, there are other methods that are useful for measuring CO_2 in the respiratory gases such as infrared spectroscopy, mass spectroscopy, Raman spectroscopy, and most recently, a colorimetric device using chemically treated filter papers (Ward & Yealy, 1998a). Colorimetric CO_2 device is used to detect the presence of CO_2 during exhalation. It has a pH-sensitive chemical indicator that changes the color from purple to yellow when the exhaled CO_2 is detected. In clinical practice, colorimetric CO_2 detector mostly finds its application during intubation to verify appropriate placement of the ETT. However, this device is not suitable for preterm infant as the color may not change regardless of the flow of exhaled gas (Williams et al., 2021). Mass and Raman spectroscopies are not applicable in the ED or out-of-hospital analysis of CO_2 (Ward & Yealy, 1998a). The most recently and widely used method is capnography that use the infrared technique. Apart from the $EtCO_2$ reading, existing commercially capnographs display the CO_2 waveform and other parameters specifically the respiratory rate (RR) and the peripheral capillary oxygen saturation (SpO_2) (Medtronic, 2022). Based on graphical representation, capnography can be classified as either time-based or volume-based capnography. With time-based capnography, the exhaled CO_2 concentration is plotted as a temporal series, whereas volume-based capnography displays the amount of exhaled CO_2 in one tidal breath (Kremeier et al., 2020). The next sections provide the overview of volumetric capnography and its limitations. And then time-based capnogram, its interpretation and possible abnormalities are discussed in-depth.

4.4.3 Volumetric capnography

Volumetric capnography (Vcap) represents the amount of CO_2 in the exhaled breath by plotting the CO_2 output on the y-axis and the expiratory tidal volume on the x-axis (Fig. 4.2). This enables breath-by-breath

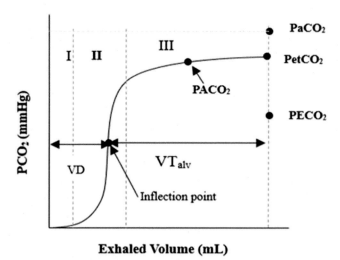

Figure 4.2 Illustration of volumetric capnogram consisting of three phases: phase I, II, and III.

quantification of anatomical dead space and alveolar dead space (Verscheure et al., 2016). Unlike time capnography, Vcap capnogram displays only the expiratory segment, which is composed of three phases specifically phase I, II and III. Phase I corresponds to the exhalation of anatomical dead space and it is free of CO_2. At this phase, approximately 10%–12% of tidal volume is exhaled. Phase II is characterized by the rapid increase in the CO_2 exhalation and 15%–18% of V_T is eliminated. The gas exhaled at phase II is a mixture of gas from the airways and the alveoli (Almeida et al., 2011). Phase III corresponds to the removal of pure alveolar gas (70%–75% of V_T). In addition, Vcap has two slopes (i.e., slope of phase II and slope of phase III) (Kremeier et al., 2020). The normal value the slope of phase II ranges from 0.36 to 0.40 mmHg/mL and this slope relates to how lung units empty at different rates of CO_2 into the main airways. On the other hand, slope of phase III is associated with the non-uniform distribution of ventilation and perfusion within lungs and its normal value is between 0.007 and 0.017 mmHg/mL (Kremeier et al., 2020). In asthmatic patients, an increased slope of phase III has been proposed to be an indicative of ventilation inhomogeneity in distal air spaces (Almeida et al., 2011). The main parameter of volumetric capnogram is the area under the curve, which corresponds to the amount of CO_2 in the breath and it reflects effective ventilation (Bhavani-Shankar et al., 1992).

Different partial pressures of CO_2 are also demonstrated: $PaCO_2$ is the partial pressure of CO_2 contained in the arterial blood; $PetCO_2$ is the end-tidal partial pressure of CO_2 (the maximum value of CO_2 at the end of exhalation); $PACO_2$ is the average value of alveolar partial pressure of CO_2, and $PECO_2$ is the mixed exhaled partial pressure of CO_2. VDaw is the airways dead space while VTalv corresponds to the alveolar tidal volume. Between VDaw and VTalv, there is an inflection point known as airways-alveolar interface (Kremeier et al., 2020).

Volumetric capnogram provides different features that are clinically significant such as airway dead space, alveolar dead space and alveolar tidal volume (Kremeier et al., 2020). Conducting airways contain CO_2-free gas known as airway dead space and it does not contribute to the gas exchange. In adult, this volume is approximately 150 mL. Alveolar dead space is consisting of alveoli that are ventilated but not perfused, and they are not involved in the gas exchange process. In a healthy person, alveolar dead space is relatively small, but it can be increased due to some pulmonary diseases. The total of VDaw and VDalv gives physiological dead space (VDphys), which is regarded as a wasted ventilation (Aung et al., 2019).

Despite of being a standard monitoring method, time-based capnography is independent of the flow rate, and does not provide information related to the volume of gas exchanged (Carlon et al., 1988). Conversely, Vcap allows the detection of the volume components of the signal, which is useful for the evaluation of the anatomical production of CO_2 and interpretation of pathological processes as well (Verscheure et al., 2016). Vcap helps to measure dead space volumes specifically, anatomical and alveolar dead space, that reveal clinical information related to the adequacy of gas exchange (de Oliveira & Moreira, 2015). However, its application to neonatal patients, particularly in newborn with stiff lungs, is limited due to different physiological and technical factors. In neonates, exhalation time is short, and the tidal volume is lower compared to adults and thus, affecting dead space measurements. The apparatus dead space mostly generated when the Vcap is connected in series with the CO_2 analyzer and pneumotach, can result in rebreathing of exhaled CO_2, which subsequently affects the accuracy of inspiratory and expiratory CO_2 measurements (Schmalisch, 2016). In Vcap, computation of dead spaces using Fletcher's method relies on subdivision of capnogram into different phases. This may be not applicable to neonates with stiff lungs. Because in this category of patients, phase II and III are not well defined and the alveolar plateau is usually absent (Schmalisch, 2016). Time and volume capnograms can both help to assess respiratory status of the patients. However, time-based capnography is the standard and adequate method for clinical applications. In fact, time capnography is simple and convenient; it allows monitoring the entire breath cycle (i.e., both inhalation and exhalation dynamics) and it is also applicable for monitoring non-intubated patients.

4.5 Time based capnogram and interpretations

The most common method of measuring CO_2 in expired air is called the infra-red absorption spectrometry. This is a more compact and less costly method of carbon dioxide measurement allowing for portability and mobility of such devices. The principle behind using infrared rays has to do with the interaction of the molecules of CO_2 in the air with the infrared ray emitted at a particular wavelength. The amount of light absorbed is directly proportional to the concentration of absorbing molecules, in this case CO_2. The CO_2 in the air will actually absorb a certain amount of the infrared ray at a particular wavelength (Ward & Yealy, 1998a). Hence, the concentration of the CO_2 can be valued by comparing the measured absorbance value with the standard value that has already been set; this value is expressed as partial pressure of CO_2 in units of mmHg (Bhavani-Shankar et al., 1992).

As mentioned in previous chapter, in earlier years, CO_2 measurement was done as discrete measurement of CO_2 concentrations, known as capnometry. This will either recognize that a certain level of CO_2 was present or provide a reading of the CO_2 concentration in the measured sample. The highest CO_2 concentration is almost always found at the end of the expiratory wave and is often referred to as the EtCO$_2$ reading. The technology of capnometry soon gave way to the latest technology of capnography, which is the continuous measurement of CO_2 multiple times each second and representing the readings into a graphical mode, the capnogram. This produces a continuous capnography waveform, which indicates how the concentration of CO_2 changes over time measured with multiple times per breath (Parker et al., 2018). In most of today's capnographs, both a numeric reading of EtCO$_2$ as well as a graphical capnogram are displayed. Appropriate analysis and interpretation of capnogram requires to understand different phases of capnogram, how a normal capnogram looks like, and possible deformations of that capnogram waveform resulting from ineffective respiratory or cardiopulmonary system or equipment defects.

4.5.1 Phases of capnogram

When a person breathes in, a stream of surrounding air is sucked into the breathing passages, passing the CO_2 sensors, which will detect the very low levels of CO_2 normally found in surrounding atmospheric air (basic science reminder — atmospheric CO_2 levels is usually around 0.04% only equivalent to 410 ppm (Babin et al., 2021), which is very low). During inspiration, this atmospheric air will be sucked into the lungs filling up the alveoli. At the end of inspiration, this air now fills all parts of the lungs from the alveoli to the upper breathing passages. As exhalation begins, the first part of the expired air that comes out (and therefore is measured first) is the air that was in the upper breathing passages. As there is no exchange of $O_2 - CO_2$ that occurs in the breathing passages, this air is essentially unchanged from the atmospheric air and thereby contains very low levels of CO_2, similar to atmospheric air. In the capnogram, this low level of recorded CO_2 is seen as part of Phase 1 of the waveform (demonstrated as AB in Fig. 4.3 below).

The $O_2 - CO_2$ exchange occurs in the alveoli. As the air from the alveoli, which is now rich in CO_2 starts coming out during exhalation, this mixes with the air in the breathing passages causing a rapid rise in the detected CO_2 levels of the capnogram (reflected in BC in Fig. 4.3). As discussed earlier in Chapter 1, in normal lungs, most alveoli drain out their air simultaneously, almost like a 'wall of alveoli air' discharging at the same time, which leads to a rapid rise in CO_2 levels, quickly reaching a high plateau. The horizontal part CD of the capnogram is often referred to as the plateau, which is the phase in which a relatively constant CO_2 concentration is detected. This is the concentration of CO_2 that is purely in the

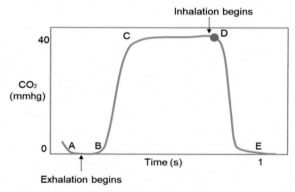

Figure 4.3 Illustration of capnographic waveform with its phases.

alveoli, where the $O_2 - CO_2$ exchange has occurred. The end part of the plateau is the highest reading of CO_2 concentration obtained, and is usually read as $EtCO_2$, in mmHg or ppm or % percentages (Malik et al., 2016; Singh et al., 2018). This marks the end of exhalation and the beginning of inhalation again, where the CO_2 reading at the sensors will again read the CO_2 concentration in incoming atmospheric air, and the waveform drops back to its original low levels (displayed as DE).

4.5.2 Understanding the normal capnogram

A capnogram is a continuous graphical display of the CO_2 concentration (mmHg) versus time (seconds). It reflects development of the respiratory condition of a patient. A normal capnogram has four phases and an end-tidal point (Fig. 4.4). Each phase expresses a section of the usual process of CO_2 elimination. The flat first phase indicates relatively CO_2-free early exhalation. As exhalation continues, the expired CO_2 increases very rapidly, and it results in the near-vertical rise in phase II. Phase III begins near the end of the normal exhalation. The end of this plateau phase is marked D-the point at which the alveolar CO_2 measurements are almost equivalent to the partial pressure of CO_2 in the arteries ($PaCO_2$). At this point, the level of sampled CO_2 is referred to as either $PetCO_2$ or $EtCO_2$—the end-tidal point. Inspiration phase (IV) begins immediately after the end-tidal

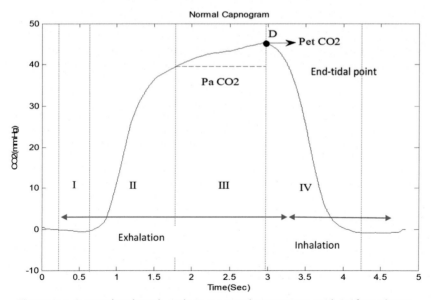

Figure 4.4 A complete breath cycle in a normal capnogram, with its four phases.

exhalation is reached. This phase is marked by rapid falling in the measured CO_2 levels to the respiratory baseline (Rhoades & Thomas, 2002). The alveolar air is measured at the end of the horizontal plateau, or the end-tidal point ($EtCO_2$), corresponding to the end of exhalation. This is usually the point with the highest CO_2 reading. In normal lungs, $EtCO_2$ values are very close to blood CO_2 values, with the former usually being just a few mmHg lower, as mentioned earlier.

In a healthy subject, the CO_2 waveform, or the capnogram, has the almost square shape. This is comprised primarily of the exhalation of CO_2-free gas or airway dead space, a rapid upstroke (indicating the "wall of alveoli air", containing a mixture of CO_2 from airways and alveoli), a horizontal plateau (indicating the constant level of CO_2 expelled from the alveoli) and a rapid downstroke (indicating the smooth flow of inspired air) (Brast et al., 2016; Kunkov et al., 2005). However, the morphology of the capnogram changes due to changes in breathing, ventilation, airway obstructions, or other breathlessness-associated conditions.

4.5.3 Exhaled CO_2 detection

While attempting to perform capnography measurement, there is a possibility that the exhaled CO_2 may not be detected by the measuring device. In this case, three main causes should be suspected: deficiency of alveolar ventilation, complete circulatory failure, and malfunction of capnograph. Differential diagnosis to be considered while assessing deficiency of alveolar ventilation includes esophageal intubation, blockage of the endotracheal tube (ETT), obstruction or disconnection of different parts of the breathing circuit. Failure to detect CO_2 secondary to complete respiratory inefficiency, can be a result of enormous pulmonary embolism, cardiac arrest, or any other situation that inhibits circulation of blood through pulmonary system normally or artificially. Finally, equipment failure can also be the reason for the CO_2 to not be detected. In this regard, sampling part of the device must checked, and the exhalation must be done actively (Ward & Yealy, 1998a).

4.5.4 Abnormalities in capnogram

The CO_2 levels measured with mainstream capnography is widely used in the clinical areas of anesthesia and critical care as a monitoring tool in ventilated patients. Despite that, sidestream capnography, in patients who are not mechanically ventilated, and breathing on their own, can be useful in different ways. The earlier discussion explained the physiology of why $EtCO_2$ by itself is not very useful. However, understanding more about the shape of the capnographic waveform, instead of just the $EtCO_2$ reading, may lead toward the development of more significant capnographic waveforms applications. The square shaped CO_2 waveform, or the capnogram, is typical of a normal waveform obtained during quiet respirations in normal lungs. It has been ascertained that CO_2 waveforms change in asthma reflect the degree of airway obstruction, and severity of disease. These changes in the capnographic waveform are primarily a reflection of the varied emptying of the millions of alveoli seen in asthma. As the disease improves with treatment, the capnographic waveform slowly reverts to a more normal shape (Surrey et al., 2018). Correspondingly, similar changes occur in other conditions of small airways narrowing like chronic obstructive pulmonary (COPD) (Long et al., 2017).

Apart from its shape, the capnogram may potentially reveal other clinically relevant findings. As an example, in acute pulmonary edema, the rapid breaths taken by the patient produce small symmetrical "mini-breath" waves, interspersed with larger waveforms that resemble the normal expiratory waveform. This is physiologically explained by the increased need for oxygen in these patients, who are forced to breathe rapidly to get O_2. This results in the lungs generally "oscillating" in and out at very high filling volumes. Occasionally, patients must produce a full expiratory cycle in order to clear the lungs and allow some of the air out. As a result of the high respiratory rates, the washout of CO_2 is unhindered and mostly peak CO_2 levels are generally low. Other features of the capnogram shape that may be relevant in clinical diseases include change of the waveform revert to baseline (which may suggest rebreathing and/or CO_2 retention), higher $EtCO_2$ trends, and inconsistent respiratory phases.

4.5.4.1 Missing alveolar plateau

Capnogram waveform displays partial pressure of carbon dioxide ($PaCO_2$) measured continuously, revealing the patient end-tidal carbon dioxide ($PetCO_2$) released at the end of each breath cycle. The partial pressure exhibited by CO_2 during and at the end of breath cycle is evaluated in capnogram to determine its abnormalities in the waveform and the patient respiration condition. When ventilation and perfusion function normally, $PetCO_2$ measurement should be 2–5 mmHg lower than the partial pressure of carbon dioxide ($PaCO_2$). This natural CO_2 gradient ($PaCO_2$- $PetCO_2$) exists between the levels of CO_2 in the artery and the end-tidal point because alveolar ventilation and perfusion are not homogenous in all areas of the lungs (Gravenstein et al., 2011). Abnormal clinical states that cause a widened gradient (>5 mmHg) includes incomplete alveolar emptying, hyperventilation and hypoperfusion of alveoli. In these conditions, the obtained $PetCO_2$ may be significantly low (Carlon et al., 1988) and subsequently not accurately correlating with the $PaCO_2$ that is measured from the patient. When expiration is not completed, the CO_2 detector monitors both the gas contained in the anatomic dead space that is CO_2-free and a portion of alveolar gas. As a result, CO_2-rich alveolar gases are only partially assessed. Therefore, the nearly horizontal plateau phase is not obtained as illustrated in Fig. 4.5, and the reported $PetCO_2$ may misinterpret the patient's true alveolar PCO_2.

Different clinical situations can cause shortening or prolongation of capnogram segments and the alteration of the $PetCO_2$ measurements (Bhavani-Shankar et al., 1992). Incomplete exhalation associated with the absence of alveolar plateau and loss of verticality of the expiratory stroke, may result from partial elimination of the alveolar air caused by the upper airway obstruction or kinked endotracheal tube (Ward & Yealy, 1998b). Furthermore, this type of CO_2 signal can also be observed while using sidestream capnograph with a slow sampling flow rate (Ward & Yealy, 1998a). When these clinical events are present, the true end-tidal point is never reached. In mechanically ventilated patient, the absence of alveolar plateau can reflect abnormal distribution of the intrapulmonary gas (Carlon et al., 1988).

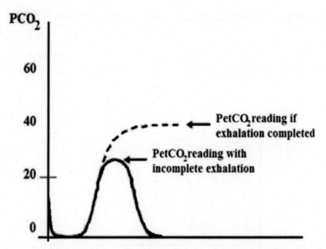

Figure 4.5 Capnogram pattern illustrating incomplete exhalation phase with decreased $PetCO_2$ reading and absence of alveolar plateau (Rhoades & Thomas, 2002).

4.5.4.2 Increasing PetCO$_2$ trend data and respiratory baseline

Another type of capnogram, as shown in Fig. 4.6. The signal has a normal shape but there is a gradual increase of the baseline and PetCO$_2$ trend data (Ward & Yealy, 1998a). The expiratory CO$_2$ concentration does not return to the actual baseline before the next respiratory cycle starts. The slight elevation of the baseline manifests rebreathing of the respiratory CO$_2$ during mechanical exhalation. In this case, the inhaled gas is a mixture of fresh and exhaled gas and pulmonary CO$_2$ is ineffectively cleared when the inspired CO$_2$ concentration increases higher than the atmospheric value (Carlon et al., 1988). This capnogram pattern may be a suggestive that the sensor is contaminated either by secretions or water vapor (Surrey et al., 2018). Consistent increase in the PetCO$_2$ trend may be associated with ineffective minute ventilation, or chronic hypercarbia (Rhoades & Thomas, 2002).

Different conditions such as ineffective respiratory flow, failure of the CO$_2$ absorber system, deficiency of expiratory valve of the circuit, and shortened expiratory time may also cause elevation in the baseline (Long et al., 2017). This type of signal could also be observed due to two main causes. First are clinical conditions that lead to elevated patient's metabolism (e.g., sepsis, thyroid storm, seizures, or fever). The second are clinical events caused by reduced alveolar ventilation effectiveness (e.g., respiratory depression, COPD, muscle paralysis, neuromuscular obstruction) (Rhoades & Thomas, 2002).

4.5.4.3 Exponential decrease and sudden decrease in PetCO$_2$

A capnogram marked by exponential decrease in PetCO$_2$, with a waveform in regular shape, is usually an indicative of cardiopulmonary system inefficiency due to rapid and remarkable rise in dead space caused by an increasing

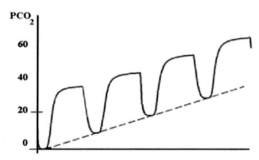

Figure 4.6 Abnormal changes in capnogram waveform. (A) Increasing PetCO$_2$ trend with gradual rise in the respiratory baseline value (Ward & Yealy, 1998a).

difference between alveolar-arterial carbon dioxide difference (Ward & Yealy, 1998b). In fact, an exponential decrease in $PetCO_2$ will occur when a patient experience extreme pulmonary hypoperfusion, cardiac arrest, or a large pulmonary embolism. Patients placed on cardiopulmonary bypass also will demonstrate a similar waveform (Parker et al., 2018). Also, a recorded capnogram can demonstrate a sudden reduction of $PetCO_2$ to low or almost zero levels as shown in Fig. 4.7 (B). A sudden decrease of $PetCO_2$ to zero or close to zero values with indistinct waveform could reflect a warning clinical event and it can result from different conditions such as esophageal intubation, improper placement of endotracheal tube (ETT), complete or partial obstruction of the ETT, and ventilator failure (Ward & Yealy, 1998b).

In patients with chronic respiratory airway diseases, capnogram pattern is distorted in a such way that the expiratory phase is enlarged than normal Fig. 4.8. Both the upstroke phase and alveolar phase are deformed and the intermediate angle that is normally between them is absent. The upstroke phase is prolonged and alveolar plateau is upward slanted (Thompson & Jaffe, 2005). This capnogram pattern can be produced when the patient has asthma, chronic obstructive pulmonary disease (COPD), or partially obstructed upper-airways or partial obstruction in the expiratory parts of equipment such as a partially kinked ETT (Ward & Yealy, 1998a).

Furthermore, various abnormal capnograms that may be recorded from patients assisted by mechanical ventilation are illustrated on Fig. 4.9, where (A) capnogram shape obtained when there is a rebreathing or a fluttering expiratory valve resulting from water condensation or fluctuation of the pressure compensation valve of the ventilator (Carlon et al., 1988; Thompson & Jaffe, 2005) and (B) asynchronies in patient-ventilator during intermittent mandatory ventilation. Asynchronies can result from insufficiency

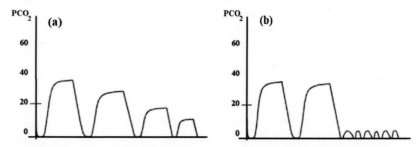

Figure 4.7 Capnogram pattern indicating exponential decrease in $PetCO_2$ (A) and sudden reduction of $PetCO_2$ to low or near zero levels (B) (Ward & Yealy, 1998b).

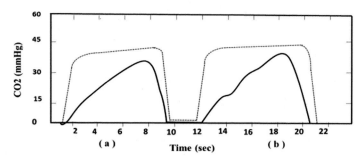

Figure 4.8 Capnogram of a patient with obstructive airway disease. (a) Upstroke phase and alveolar plateau are undefined compared to the normal capnogram (dotted lines); (b) An increased tidal volume with extended expiratory phase indicates PaO$_2$.

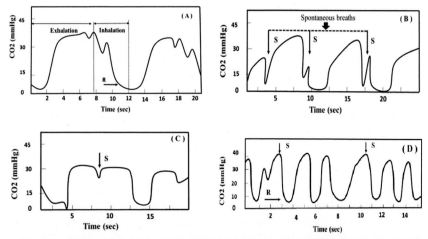

Figure 4.9 Various abnormal capnograms that may be recorded from mechanically ventilated patients in different conditions: Fluttering expiratory valve and the arrow R indicates rebreathing; Patient ventilator asynchrony during assisted ventilation (B); An abnormal capnogram in which the arrow S indicates insufficient spontaneous inspiratory effort that failed to trigger the ventilator (C); Capnogram representing weaning failure (D). The signal has irregular breath cycles in which the arrow R indicates rebreathing and S shows spontaneous breaths.

of ventilator assistance to meet the patient's demand (De Haro et al., 2019); (C) Capnogram indicating a situation in which the respiratory effort is very insufficient and cannot trigger ventilator. This situation requires to adjust triggering sensitivity; (D) a capnogram pattern showing weaning failure (i.e., the patient fails to pass a spontaneous breathing trial when ventilator assistance is gradually decreased). There is a disordered, fast breathing associated with rebreathing (R). The arrows (S) indicate spontaneous breaths at the time of mandatory breaths (Thompson & Jaffe, 2005).

4.6 Clinical interpretation of capnogram

Although much more research is needed in this field of interest, the potential uses of capnogram analysis is significant. Many clinical situations present similarly; but the treatments may be very different. Acute breathlessness is one such situation. Asthma, exacerbations of COPD and acute pulmonary oedema with acute heart failure all present with acute onset of breathlessness. Very often, all have wheeze on clinical examination too. Sometimes it even co-exist. The treatments are markedly different; and treatment for asthma and COPD given to a patient with acute heart failure will be detrimental. Therefore, it is of prime importance to be able to diagnose and differentiate these conditions. Capnogram analyses can help with that. Also, as the waveform changes with treatment, it will help clinicians monitor response to treatments continuously and act accordingly with greater precision. The next chapters discuss on analysis of capnogram waveform. Apart from diagnosing and differentiating presentations of acute breathlessness, recognition of different waveforms may provide insight into how disease processes may be occurring by seeing their impacts on the waveform. This could potentially provide earlier identification and complications of diseases or medications and could allow physicians to tailor treatment more specifically for certain patients particularly asthma.

References

Albert, R. K., Spiro, S. G., & Jett, J. R. (2008). *Clinical respiratory medicine*. Elsevier Health Sciences.

Almeida, C. C., Almeida-Júnior, A. A., Ribeiro, M.Â. G., Nolasco-Silva, M. T., & Ribeiro, J. D. (2011). Volumetric capnography to detect ventilation inhomogeneity in children and adolescents with controlled persistent asthma. *Jornal de pediatria, 87,* 163−168.

Aminiahidashti, H., Shafiee, S., Kiasari, A. Z., & Sazgar, M. (2018). Applications of end-tidal carbon dioxide (ETCO2) monitoring in emergency department; a narrative review. *Emergency, 6*(1).

Aung, H., Sivakumar, A., Gholami, S., Venkateswaran, S., & Gorain, B. (2019). An overview of the anatomy and physiology of the lung. *Nanotechnology-Based Targeted Drug Delivery Systems for Lung Cancer,* 1−20.

Babin, A., Vaneeckhaute, C., & Iliuta, M. C. (2021). Potential and challenges of bioenergy with carbon capture and storage as a carbon-negative energy source: A review. *Biomass and Bioenergy, 146,* 105968.

Bhavani-Shankar, K., Moseley, H., Kumar, A., & Delph, Y. (1992). Capnometry and anaesthesia. *Canadian Journal of Anaesthesia, 39*(6), 617−632.

Brast, S., Bland, E., Jones-Hooker, C., Long, M., & Green, K. (2016). Capnography for the radiology and imaging nurse: A primer. *Journal of Radiology Nursing, 35*(3), 173−190.

Calcagno, M., Kahleova, H., Alwarith, J., Burgess, N. N., Flores, R. A., Busta, M. L., & Barnard, N. D. (2019). The thermic effect of food: A review. *Journal of the American College of Nutrition, 38*(6), 547–551.

Carlon, G. C., Ray, C., Jr., Miodownik, S., Kopec, I., & Groeger, J. S. (1988). Capnography in mechanically ventilated patients. *Critical Care Medicine, 16*(5), 550–556.

Castro, D., Patil, S. M., & Keenaghan, M. (2021). *Arterial blood gas*. StatPearls.

Chambers, D., Huang, C., & Matthews, G. (2019). *Basic physiology for anaesthetists*. Cambridge University Press.

De Haro, C., Ochagavia, A., López-Aguilar, J., Fernandez-Gonzalo, S., Navarra-Ventura, G., Magrans, R., Montanyà, J., & Blanch, L. (2019). Patient-ventilator asynchronies during mechanical ventilation: Current knowledge and research priorities. *Intensive Care Medicine Experimental, 7*(1), 1–14.

de Oliveira, & Moreira, M. M. (2015). Capnography: a feasible tool in clinical and experimental settings. *Respiratory Care, 60*(11), 1711–1713.

Deenstra, D. D., van Helvoort, H. A., Djamin, R. S., van Zelst, C., In't Veen, J. C., Antons, J. C., Spruit, M. A., & van't Hul, A. J. (2021). Prevalence of hyperventilation in patients with asthma. *Journal of Asthma*, 1–8.

Dicembrino, M., Alejandra Barbieri, I., Pereyra, C., & Leske, V. (2021). End-tidal CO_2 and transcutaneous CO_2: Are we ready to replace arterial CO_2 in awake children? *Pediatric Pulmonology, 56*(2), 486–494.

Gravenstein, J. S., Jaffe, M. B., Gravenstein, N., & Paulus, D. A. (2011). *Capnography*. Cambridge University Press.

Guthrie, B. D., Adler, M. D., & Powell, E. C. (2007). End-tidal carbon dioxide measurements in children with acute asthma. *Academic Emergency Medicine, 14*(12), 1135–1140.

van't Hul, A. J., Deenstra, D. D., Djamin, R. S., Antons, J. C., & van Helvoort, H. A. (2018). Hypocapnia correction as a working mechanism for breathing retraining in asthma. *The Lancet Respiratory Medicine, 6*(4), e14.

Kremeier, P., Böhm, S. H., & Tusman, G. (2020). Clinical use of volumetric capnography in mechanically ventilated patients. *Journal of Clinical Monitoring and Computing, 34*(1), 7–16.

Kunkov, S., Pinedo, V., Silver, E. J., & Crain, E. F. (2005). Predicting the need for hospitalization in acute childhood asthma using end-tidal capnography. *Pediatric Emergency Care, 21*(9), 574–577.

Langhan, M. L., Zonfrillo, M. R., & Spiro, D. M. (2008). Quantitative end-tidal carbon dioxide in acute exacerbations of asthma. *The Journal of Pediatrics, 152*(6), 829–832.

Long, B., Koyfman, A., & Vivirito, M. A. (2017). Capnography in the emergency department: A review of uses, waveforms, and limitations. *The Journal of Emergency Medicine, 53*(6), 829–842.

Malik, S. A., Singh, O. P., Nurifhan, A., & Malarvili, M. (2016). Portable respiratory CO_2 monitoring device for early screening of asthma. In *Proc. ACEC*.

Medtronic. (2022). *Capnostream™ 20p bedside monitor with Apnea-Sat Alert algorithm*. Retrieved 04 April from https://www.medtronic.com/covidien/en-us/products/capnography/capnostream-20p-bedside-patient-monitor.html.

Mehta, J. H., Williams, G. W., Harvey, B. C., Grewal, N. K., & George, E. E. (2017). The relationship between minute ventilation and end tidal CO_2 in intubated and spontaneously breathing patients undergoing procedural sedation. *PLoS One, 12*(6), e0180187.

Neligan, P. J., & O'Donoghue, R. (2010). 56 how should acid-base disorders be diagnosed and managed? In Clifford S. Deutschman, & Patrick J. Neligan (Eds.), *Evidence-based practice of critical care* (pp. 389–396). Philadelphia: WB Saunders.

Parker, W., Estrich, C. G., Abt, E., Carrasco-Labra, A., Waugh, J. B., Conway, A., Lipman, R. D., & Araujo, M. W. (2018). Benefits and harms of capnography during procedures involving moderate sedation: A rapid review and meta-analysis. *The Journal of the American Dental Association, 149*(1), 38.e32-50.e32.

Patel, S., & Sharma, S. (2021). Respiratory acidosis. In *StatPearls [Internet]*. StatPearls Publishing.

Pizano, A., Calvacci, P., Giron, F., & Cordovez, J. (2019). Ventilation-perfusion ratio: A mathematical approach for gas exchange in the lungs. *The FASEB Journal, 33*(S1), 600.609–600.609.

Powers, K. A., & Dhamoon, A. S. (2019). *Physiology, pulmonary ventilation and perfusion*.

Qureshi, S. M., & Mustafa, R. (2018). Measurement of respiratory function: Gas exchange and its clinical applications. *Anaesthesia & Intensive Care Medicine, 19*(2), 65–71.

Raimondi, G. A., Gonzalez, S., Zaltsman, J., Menga, G., & Adrogué, H. J. (2013). Acid–base patterns in acute severe asthma. *Journal of Asthma, 50*(10), 1062–1068.

Rhoades, C., & Thomas, F. (2002). Capnography: Beyond the numbers. *Air Medical Journal, 21*(2), 43–48.

Robertson, H. T. (2015). Dead space: The physiology of wasted ventilation. *European Respiratory Journal, 45*(6), 1704–1716.

Ruiz, J. M., De Gauna, S. R., González-Otero, D. M., Daya, M., Russell, J. K., Gutierrez, J., & Leturiondo, M. (2016). Relationship between $EtCO_2$ and quality-parameters during cardiopulmonary resuscitation. In *2016 Computing in Cardiology Conference (CinC)*.

Schmalisch, G. (2016). Current methodological and technical limitations of time and volumetric capnography in newborns. *Biomedical Engineering Online, 15*(1), 1–13.

Selby, S. T., Abramo, T., & Hobart-Porter, N. (2018). An update on end-tidal CO_2 monitoring. *Pediatric Emergency Care, 34*(12), 888–892.

Singh, O. P., Howe, T. A., & Malarvili, M. (2018). Real-time human respiration carbon dioxide measurement device for cardiorespiratory assessment. *Journal of Breath Research, 12*(2), 026003.

Stine, C. N., Koch, J., Brown, L. S., Chalak, L., Kapadia, V., & Wyckoff, M. H. (2019). Quantitative end-tidal CO_2 can predict increase in heart rate during infant cardiopulmonary resuscitation. *Heliyon, 5*(6), e01871.

Surrey, A., Lambert, A., & Evans, D. (2018). End tidal capnography in the emergency department. *Emergency Medicine, 50*, 1–9.

Taytard, J., Lacin, F., Nguyen, T. L. T., Boizeau, P., Alberti, C., & Beydon, N. (2020). Children with uncontrolled asthma and significant reversibility might show hypoxaemia. *European Journal of Pediatrics, 179*(6), 999–1005.

Thompson, J. E., & Jaffe, M. B. (2005). Capnographic waveforms in the mechanically ventilated patient. *Respiratory Care, 50*(1), 100–109.

Tsakiris, T. S., Konstantopoulos, A. I., & Bourdas, D. I. (2021). The role of CO_2 on respiration and metabolism during hypercapnic and normocapnic recovery from exercise. *Research Quarterly for Exercise and Sport, 92*(3), 537–548.

Van den Elshout, F., Van Herwaarden, C., & Folgering, H. (1991). Effects of hypercapnia and hypocapnia on respiratory resistance in normal and asthmatic subjects. *Thorax, 46*(1), 28–32.

Verscheure, S., Massion, P. B., Verschuren, F., Damas, P., & Magder, S. (2016). Volumetric capnography: Lessons from the past and current clinical applications. *Critical Care, 20*(1), 1–9.

Wagner, P. D. (2015). The physiological basis of pulmonary gas exchange: Implications for clinical interpretation of arterial blood gases. *European Respiratory Journal, 45*(1), 227–243.

Ward, K. R., & Yealy, D. M. (1998a). End-tidal carbon dioxide monitoring in emergency medicine, Part 1: Basic principles. *Academic Emergency Medicine, 5*(6), 628–636.

Ward, K. R., & Yealy, D. M. (1998b). End-tidal carbon dioxide monitoring in emergency medicine, part 2: Clinical applications. *Academic Emergency Medicine, 5*(6), 637–646.

Williams, E., Dassios, T., & Greenough, A. (2021). Carbon dioxide monitoring in the newborn infant. *Pediatric Pulmonology, 56*(10), 3148–3156.

Wolfson, M. R., & Shaffer, T. H. (2004). Respiratory physiology: Structure, function, and integrative responses to intervention with special emphasis on the ventilatory pump. In *Cardiopulmonary physical therapy* (pp. 39–81). Elsevier.

CHAPTER FIVE

Analysis of capnogram using signal processing techniques

Capnography is a rapid and noninvasive method of continuously monitoring the concentration of carbon dioxide (CO_2) in exhaled gases and it has become an important tool to aid clinical diagnosis of respiratory status (Böhm et al., 2020). In this chapter, different studies carried out on capnogram waveform and various signal processing techniques for the analysis of capnogram are discussed. The main goal in this chapter is to quantify the state of the capnogram during the presence and absence of asthma. In other words, numerous features of capnogram that can be used to differentiate between asthma and nonasthma conditions are described. Those features can be used to design a robust automatic algorithm to detect asthma and nonasthmatic conditions.

5.1 Early studies on expired CO_2 signal (1960–90)

From 1960s, early studies analyzed the alteration of carbon dioxide (CO_2) concentration in normal subjects and patients with abnormal respiratory conditions. Kelsey et al. (1962) studied the modification of the CO_2 curve in pulmonary emphysema patients. Patient exhalations have been recorded using an infrared gas analyzer equipped with breath-through type cell and a strip chart recorder (Tolnai et al., 2018). The contour of the CO_2 expiratory curve was assumed to vary with the severity of the disease. Carbon dioxide retention, apparently absent in the mild emphysema patient, occurred as the severity of the disease increases. It was noted that as the patients improve due to the treatment, the CO_2 expiratory curve tends to become more normal in shape (Kelsey et al., 1962). Van Meerten (1971) experimentally evaluated the exponential nature of the expiratory gas concentration curves using mathematical model. The minimum radius R_{min} of the curvature of the area between the steep slightly S-shaped transition phase and the horizontal alveolar phase was measured. The R_{min} has been then suggested as a diagnostic criterion for the evaluation of centrilobular emphysema and nonuniformity of ventilation (Van Meerten, 1971). In 1976, Vargha analyzed the CO_2 concentration curve with the aid of

Systems and Signal Processing of Capnography as a Diagnostic Tool for Asthma Assessment
ISBN: 978-0-323-85747-5
https://doi.org/10.1016/B978-0-323-85747-5.00010-3

geometric method. By subdividing the curve into different segments using six auxiliary lines, the T_2/T_1 ratio was determined. The gamma (γ) angle was measured using two directional lines on the second part and the third part of the expiratory phase. The lessening in the T_2/T_1 ratio and the increase in gamma angle was closely associated with the increase of the severity of obstructive ventilatory disturbance. However, γ angle could not then be computed for few capnograms (Vargha, 1976).

5.2 Capnogram features

The varying airway obstruction result in the change of the CO_2 signal, which is shown by the change in uprightness of the ascending phase (II), opening of the angle α, shortening and inclination of Phase III (You et al., 1994). Analysis of this deformation using capnogram features can give critical information for assessing human respiratory health. Several time and frequency domain features such as end–tidal carbon–dioxide partial pressure (PetCO$_2$), respiratory rate (RR), time spent at PetCO$_2$, duration for which the CO_2 partial pressure remains at its maximum value, Hjorth parameters, exhalation duration, end exhalation slope (α angle), area ratio, slope ratio (SR), power spectral density (PSD), linear predictive coding, and wavelet coefficients have been extracted and proposed as the indicators for cardiopulmonary diseases, specifically asthma (El–Badawy et al., 2020; Mieloszyk et al., 2014; Singh, Palaniappan, et al., 2018). Table 5.1 illustrates an overview of capnogram time and frequency domain features, which have been proposed for differentiating asthmatic conditions.

5.2.1 Time domain features of capnogram

Later in the 1990s, the abnormal shape of the CO_2 signal was analyzed more closely and capnographic indices were mostly compared with spirometry or peak flow meter parameters. For asthmatic patients, the most significant change in the capnogram pattern, is reflected by the increase in the slope of the alveolar plateau as a result of the asynchronous alveolar emptying (Kunkov et al., 2005). In addition, various factors including the airway resistance, alteration in the cardiac output and CO_2 production, have also effect on the slope of phase III (Gravenstein et al., 2011). The morphology of capnogram depends on a continuous removal of alveolar gas. Therefore, pathologies affecting phase II also affect the alveolar plateau (Gravenstein et al., 2011). Phase II and phase III are among the indices that are mostly reported in the literature for quantitative analysis of capnogram waveform. You et al.

Table 5.1 Illustration of studies conducted on capnogram time and frequency domain features for characterization of asthma and nonasthma conditions.

Authors	Features (Time/ frequency)	Performance assessment (AUC (0.xx), Accuracy %)	Cardiopulmonary diseases
Meyer et al. (1990)	A slope of alveolar Plateau/Nil	—	Reveals the exchange of gasses during respiration
You et al. (1994)	S1, S2, S3, SR, AR, SD1, SD2 and SD3/Nil	Correlation with spirometer feature	Asthma
Yaron et al. (1996)	$\frac{dCO_2}{dt}$/Nil	$r = 0.84$, $P < .001$	Asthma
Kunkov et al. (2005)	EtCO$_2$ ratio/Nil	—	Asthma
Guthrie et al. (2007)	EtCO$_2$/Nil	—	Asthma
Langhan et al. (2008)	EtCO$_2$/Nil	—	Asthma
Hisamuddin et al. (2009)	A slope of alveolar phase, α angle/ Nil	—	Asthma
Kean & Malarvili (2009)	A1 and A2 (area), an Area ratio (AR), S1 and S2 (Slope), SR (slope ratio), AR (area ratio), α angle, HP1 and HP2 (activity), HP1 and HP2 (mobility)/Nil	Slope ratio (SR) = 0.92 Activity = 0.90	Asthma
Howe et al. (2011)	Slope of Phase II & III, and α angle/ Nil	—	Asthma
Kazemi et al. (2013)	Nil/PSD, LPC coefficient	Asthma prognostic index, Accuracy: 95.65%; error rate: 4.34%	Asthma
Betancourt et al. (2014)	Nil/Wavelet coefficients	Asthma prognostic index, Accuracy:96.52%; error rate: 3.48%;	Asthma severity

(Continued)

Table 5.1 Illustration of studies conducted on capnogram time and frequency domain features for characterization of asthma and nonasthma conditions.—cont'd

Authors	Features (Time/frequency)	Performance assessment (AUC (0.xx), Accuracy %)	Cardiopulmonary diseases
		sensitivity: 100%; specificity: 88.57%; execution time: 0.3 feature estimation time- 0.35s	
Malik et al. (2016)	CO_2 concentration/ Nil	Early screening of asthma, Sensitivity: 80%; specificity: 90%	Asthma
Hong et al. (2018)	CO_2 concentration/ Nil	—	Asthma
Singh et al. (2018)	AR1, AR2, AR3, AR4, AR1+AR2, derivative (D)/Nil	Accuracy: 94.52%; error rate: 5.47%; sensitivity: 97.67%; specificity: 90.0%; execution time: 0.03 s average feature estimation time = 0.24 s	Asthma

HP, Hjorth parameter; LPC, linear predictive coding; PSD, power spectral density; S, slope; SR, slope ratio; AR1, AR2, AR3, AR4, AR1+AR2, and D are the area of upward expiratory, downward inspiratory, absolute expiratory, complete breath cycle, and sum of area of upward expiratory and downward inspiratory phase, and derivative of whole expiratory phase, respectively.

(1994) have investigated eight indices (S1, S2, S3, SR, AR, SD1, SD2 and SD3) from capnogram and all of them correlated with spirometric parameter FEV1 as a percentage of predicted. However, the angle between the ascending phase (E2) and the alveolar plateau was found to have the highest association (E3) (Fig. 5.1). Generally, all indices were found to be significant to understand the level of airways obstruction in asthmatic patients, However, those indices were computed manually (You et al., 1994).

Another study has evaluated changes in the alveolar plateau phase (dCO_2/dt) of five regular expiratory phases recorded from adult asthmatic patients. Compared to the normal subjects, the dCO_2/dt was higher in asthmatic patients, and it was strongly correlated with the PEFR. The dCO_2/dt

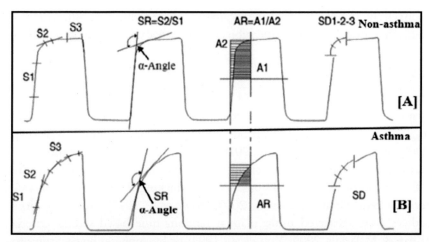

Figure 5.1 Illustration of eight indices extracted from capnograms of nonasthmatic subject (A) and asthmatic patient (B), respectively. S1, S2, and S3 represent the slopes; SR is the slope ratio; A1 and A2 are areas and AR is area ratio; SD indicates double derivatives (You et al., 1994).

was proposed to be a useful index for evaluating the therapy of acute asthma as it was reduced after taking the inhaled β-agonists therapy treatment (Yaron et al., 1996). However, computation of this parameter has been done manually, which is time consuming and measurement errors can be produced.

Furthermore, the intermediate angle (α) and slopes of phase II and III have demonstrated good performance to detect changes in the airways due to treatment (Hisamuddin et al., 2009). Howe et al. (2011) verified the feasibility of those three features. The angles were computed using the observed gradient of phase II and phase III, and the slopes were calculated from the segments of the capnogram using linear trendline analysis. Before treatment, phase II, phase III, and angle gradient values in terms of mean were (2.61, 0.44, and 134.36, respectively), but posttreatment values were (2.74, 0.23 and 123.27, respectively). The slope of phase II ($P = 0.35$) showed a minor shift. While the slope of phase III and the angle of phase III both decreased significantly with p< 0.001 for both. The insignificant change in phase II's slope could be attributed to the difficulties in picking the phase II's starting point (Howe et al., 2011). In 2018, Singh et al. proposed six features to differentiate nonasthmatic subjects from asthmatic patients by computing areas (*ARi*) and derivative of capnogram (dCO$_2$/dt). The mean value of areas was higher in asthmatic patients compared to the nonasthmatic subjects.

Conversely, the dCO_2/dt of the expiratory phase reduced in asthmatic patients than in the nonasthmatic subjects (Singh, Palaniappan, et al., 2018).

5.2.2 Hjorth parameters

Previous studies on the analysis of capnogram waveform have favored the shape indices such as slopes of different phases, slope ratio, intermediate angle (i.e., alpha angle) and various areas of the curve (Kean & Malarvili, 2009; Singh, Palaniappan, et al., 2018; You et al., 1994). Because those features can provide significant characteristic changes useful to discriminate asthma and nonasthma conditions as well as tracking changes before and after treatment (Hisamuddin et al., 2009; Howe et al., 2011). However, those indices were extracted using manual random time-based setting criteria (Kean & Malarvili, 2009; You et al., 1994) or by applying linear fitting to the capnogram (Howe et al., 2011). In fact, getting simple linear data sets in real time is quite difficult due to the uneven removal of CO_2 samples from the alveoli. Therefore, exploration of slopes of capnogram waveform is still an open research problem. Besides eight features introduced in (You et al., 1994), Hjorth parameters have been proposed as alternative means of analyzing capnogram waveform since these parameters involve the slopes of the curve (Kean & Malarvili, 2009).

Hjorth's parameters have been introduced in 1970 by Hjorth for quantitative analysis of the electroencephalogram (EEG) trace in the time domain (Hjorth, 1970). These parameters are based on the standard deviations of the amplitude of the signal, and its derivatives (i.e., the first and the second derivative) (Cocconcelli et al., 2022; Hjorth, 1970). Hjorth parameters is a set of three parameters specifically activity, mobility, and complexity (Hjorth, 1970). These parameters provide information related to the amplitude and the slopes of signal and the similarity between shape of the signal under analysis and that of a pure sine wave. Activity provides a measure of the variance of the amplitude of the signal, and it reflects the signal's variability with respect to its mean value (El–Badawy et al., 2022). When the value of the variance is relatively high, it is an indication that the amplitude of the signal is widely spread around the average value. Whereas a small variance reflects the opposite. The variance of a CO_2 signal $x(n)$ with data length N, is expressed by Eq. (5.1) (El–Badawy et al., 2022):

$$\sigma_x^2 = \frac{1}{N} \sum_{n=0}^{N-1} (x(n) - \bar{x})^2 \tag{5.1}$$

The second Hjorth parameter is mobility, and it corresponds to the ratio between the standard deviations of the first derivative of the signal and that of the signal (Oh et al., 2014). These variances quantities are equally influenced by the mean amplitude. Thus, the ratio will be dependent on the shape of the signal only and in a such a way that it measures the relative average slope (Hjorth, 1970). In the frequency domain analysis, mobility is considered as a measure of the standard deviation of the power spectrum of the signal. It means that a signal with a widely spread power spectrum has a higher mobility as compared to the signal with focused frequency content in a narrow range (El-Badawy et al., 2022). Mobility is defined as (Eq. 5.2) (El-Badawy et al., 2022):

$$\text{Mobility}[x(n)] = \frac{\sigma_{x'}}{\sigma_x} \tag{5.2}$$

where $x'(n)$, the first derivative of a CO_2 signal and it is calculated as:

$$x'(n) = \frac{1}{T_s}[x(n+1) - x(n)] \tag{5.3}$$

T_s is the sampling time interval, and it is equal to the inverse of the sampling frequency: $T_s = \frac{1}{f_s}$.

Complexity is dimensionless quantity, and it reflects the similarity between the shape of the signal under analysis and the shape of a sine wave. When the signal's morphology gets more identical to a pure sine wave, the complexity value converges to one. The minimum value of the complexity is one and this value corresponds to the complexity of a pure sinusoidal waveform in the time domain (Hjorth, 1970; Oh et al., 2014). Complexity is computed as the ratio between mobility of the first derivative of the signal and the mobility of the signal itself (Eq. 5.4) (El-Badawy et al., 2022).

$$\text{Complexity } [x(n)] = \frac{\text{mobility } [x'(n)]}{\text{mobility } [x(n)]} = \frac{\sigma_{x''}/\sigma_{x'}}{\sigma_{x'}/\sigma_x} \tag{5.4}$$

This is equivalent to,

$$\text{Complexity } [x(n)] = \frac{\sigma_{x''} \times \sigma_x}{(\sigma_{x'})^2} \tag{5.5}$$

where $x''(n)$ represents the second derivative of a CO_2 signal, and it is calculated in the same way as Eq. (5.3) (El-Badawy et al., 2022).

$$x''(n) = \frac{1}{T_s}[x'(n+1) - x'(n)] \tag{5.6}$$

Fig. 5.2 denotes the transformation of a capnogram cycle into the derived Hjorth parameters. Fig. 5.2A illustrates the capnogram waveform of a non-asthma patient. The variation in the activity is illustrated as shown in Fig. 5.2B. Fig. 5.2C denotes the differences in the mobility as the capnogram waveform were computed for its first derivate of the signal. The variation of the waveform complexity is shown in Fig. 5.2D.

In (Kean & Malarvili, 2009) Hjorth parameters and features reported in (You et al., 1994) were applied to differentiate asthmatic from nonasthmatic capnograms. Hjorth parameters (HP1) were extracted from the entire recorded signal and from 75% of the expiratory segment (HP2) starting at the beginning of the upstroke. Compared to other features, HP2-mobility

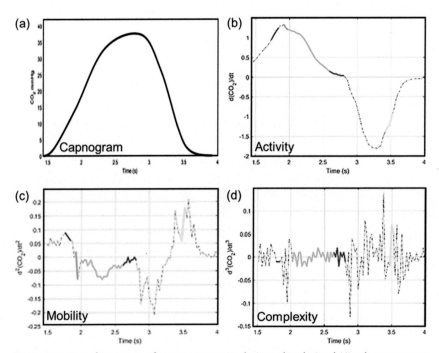

Figure 5.2 Transformation of a capnogram cycle into the derived Hjorth parameters denotes the transformation of a capnogram cycle into the derived Hjorth parameters. (A) illustrates the capnogram waveform of a non-asthma patient. The variation in the activity is illustrated as shown in (B). (C) denotes the differences in the mobility as the capnogram waveform were computed for its first derivate of the signal. The variation of the waveform complexity is shown in (D).

and the slope ratio demonstrated higher discriminatory potential (Kean & Malarvili, 2009). However, physiological implication of these statistical differences requires further investigation. In addition, generalization of their findings requires a bigger number of sample populations. Hjorth activity was highly significant while differentiating asthma and nonasthmatic conditions. In (El-Badawy et al., 2022), Hjorth mobility and the mean absolute deviation (MAD) were higher relevant than Hjorth activity and complexity for differentiating regular from irregular capnogram (El-Badawy et al., 2022). Hjorth complexity was not a suitable feature since it was almost equal in regular and irregular capnograms. High complexity values (almost five time the complexity of a sinusoidal waveform) obtained for both classes can be an indication that both regular and irregular capnogram segments are dissimilar to the sine wave (El-Badawy et al., 2022). Those three Hjorth parameters will together be used to characterize capnogram patterns based on the amplitude, time scale and complexity.

5.2.3 Entropy measures

Different features have been extracted from capnogram and proposed as indicators of respiratory conditions especially in asthmatic patients. Capnogram indices (i.e., slopes and angles) were extensively explored. However, those indices may not always be clearly defined due to the deformation of capnogram waveform in case of abnormal respiratory conditions. Thus, finding significant features that can quantify dynamics of respiratory patterns will enhance the assessment of respiratory conditions.

In recent years, several studies have been carried on the properties of complexity and regularity behaviors of physiological systems (Kapidžić et al., 2014). Biological systems are characterized by complex dynamics (Borowska, 2015) resulting from nonlinear interactions among different subunits of the system when they are operating under nonequilibrium conditions (Thamrin et al., 2016). Respiratory system exhibits complexity behavior both in health and disease conditions (Thamrin et al., 2016). Generally, complex fractals in the respiratory system includes the airways and vascular trees that are formed at the time of embryonic development (Glenny, 2011). Signals moving through those complex structures manifest complex temporal behavior (Suki, 2002). Abnormal conditions including diseases lead to the dynamic changes in structure and function of the lungs. Asthma and chronic obstructive pulmonary disease (COPD) are chronic diseases that are typically complex, influenced by different factors from external, environmental, and internal inflammatory stimuli. In addition, symptoms of those diseases vary over time. For example, there is a fluctuation in temporal behavior of asthma over time due to constant variation of stimuli (Thamrin et al., 2016). The new studies are greatly increasing from application of newly and sophisticated classical time series examining methods on different recorded physiological signals (Kapidžić et al., 2014). Nonlinear parameters have been developed to quantify regularity/complexity of time series, which leads to higher sensitivity for detecting variations in underlying physiological regulatory mechanisms (Kapidžić et al., 2014). Entropy has been considered as a suitable complexity measure for analyzing time series recorded from biological systems including brain, heart, muscles, or the uterus (Borowska, 2015).

There are numerous types of entropy measures that are used for the estimation of complexity in physiological time series including approximate entropy (*ApEn*) (Pincus, 1991), sample entropy (*SampEn*) (Richman & Moorman, 2000), permutation entropy (Ferlazzo et al., 2014), multiscale entropy (MSE) MSE) (Hadoush et al., 2019), and fuzzy entropy (*FuzzyEn*) (Xiang et al., 2015). Approximate Entropy and Sample Entropy are two algorithms defined with the intention of determining the regularity of time — series by considering the existence of patterns (Delgado-Bonal & Marshak, 2019). Approximate entropy (*ApEn*) and sample entropy (*SampEn*) enable the estimation of the randomness of a series of data even when the source generating the dataset is not previously known. Therefore, those algorithms are extensively applied in different research fields, however, they have initially developed for physiological applications (Delgado-Bonal & Marshak, 2019). The details about approximate entropy and their application for the analysis of respiratory patterns are discussed in the next sections.

5.2.3.1 Approximate entropy

Approximate entropy (*ApEn*) is a statistic that has been introduced by Pincus for quantifying regularity and complexity in time data series (Delgado-Bonal & Marshak, 2019). *ApEn* was initially applied to various physiological and clinical cardiovascular time-series (Pincus, 1991). Later, the technique was applied in other fields and was found useful to discriminate order in data series produced by stochastic and deterministic systems. *ApEn* is a parameter that measures regularity in the system by considering the existence of similar patterns within the series of data. The low *ApEn* values means that there is a high number of patterns that repeat themselves throughout of the series. This kind of system is very persistent and predictive. On the other hand, the larger *ApEn* values reflect a low number of patterns that are similar and thus the data in that series are independent each other (Delgado-Bonal & Marshak, 2019).

Computation of the *ApEn* requires first to specify two input parameters, *m* and *r* where *m* is the embedding dimension or length of sequence to be compared (a run) and it corresponds to the length of the template of the different vector comparisons and *r* is the noise filter and it also the tolerance for accepting matches. The *ApEn* algorithm require also uniformly sampled time series (Pincus & Goldberger, 1994). *ApEn* measures the logarithmic probability that nearby patterns that are close for *m* observations remain close on next incremental comparisons (Pincus, 1991). A greater likelihood of remaining close (i.e., high regularity) results in smaller approximate entropy values, and on contrary, random data (high irregularity) gives higher *ApEn* values (Caldirola et al., 2004). For a time series of N points, $\{u(i), 1 \leq i \leq N\}$, a positive integer m, with $m \leq N$ and a nonnegative real number r, the algorithm for determining the Approximate entropy of a sequence can be summarized as follows (Pincus, 1991):

1. Form a sequence of vectors $x(1)$, $x(2)$, ...$x(N-m+1)$, where the template

$$
\begin{aligned}
x(i) &= \{[u(i),\ u(i+1),\ ...,\ u(i+m-1)]\} \text{ and} \\
x(j) &= \{[u(j),\ u(j+1),\ ...,\ u(j+m-1)]\}
\end{aligned}
\tag{5.7}
$$

2. Calculate the distance between them as

$$
d\,[x(i), x(j)] = \max{}_{k=1,2,...,m}\,(|u(i+k-1) - u(j+k-1)|).
\tag{5.8}
$$

3. Calculate the value $C_i^m(r)$ = number of $j \leq N - m + 1$ such that in Eq 5.9. The numerator of C_i^m counts, within the resolution r, the number of blocks of consecutive values of length m, which are similar to a given block.

$$
d\,[x(i), x(j)] = (\leq r)/(N - m + 1)
\tag{5.9}
$$

4. Compute the function $\Phi^m(r)$, the average of the natural logarithms of the functions $C_i^m(r)$ as:

$$
\Phi^m(r) = \frac{1}{N - m + 1} \sum_{i=1}^{N-M+1} \log C_i^m(r)
\tag{5.10}
$$

5. Define

$$
ApEn(m, r, N)(u) = \Phi^m(r) - \Phi^{m+1}(r), \text{ with } m \geq 1.
\tag{5.11}
$$

$ApEn(m, r, N)(u)$ measures the logarithmic frequency with which the templates of length m that are close stay close together for the next position $(m + 1)$, or put differently, the negative value of $ApEn$ is defined as:

$-ApEn(m, r, N)(u) = \Phi^{m+1}(r) - \Phi^m(r) =$ average over i of the logarithm (conditional probability of $|u(j + m) - u(i + m)| \leq r$, if it is verified that $|u(j + k) - u(i + k)| \leq r$ for $k = 0, 1, \ldots m - 1$. $ApEn(m, r, N)(u)$ is the statistical estimator of the parameter $ApEn(m, r)$:

$$ApEn(m, r) = \lim_{N \to \infty} \left[\Phi^m(r) - \Phi^{m+1}(r) \right]. \tag{5.12}$$

5.2.3.2 Selection of input parameters

$ApEn$ is a function of m and r, so the choice of those parameters must be done carefully as they can significantly affect the value of $ApEn$. The most recommended values for m are $m = 2$ or three and r must be large enough so that in most of the specified vectors, a sufficient number of sequences of x-vectors within a distance r can be obtained, thus appropriate estimates of conditional probabilities can be ensured. The recommended r values are generally in the range of $0.1-0.25$ standard deviation of the series of data under analysis especially for slow-dynamics systems. However, the result is determined by the exact number of r selected. When the value of r decreases, the $ApEn$ (m,r) increase as $\log(2r)$, thus the $ApEn$ (m,r) can significantly change depending of the choice of r, which subsequentially affects the value of the statistic $ApEn$ (m, r, N) (Delgado-Bonal & Marshak, 2019).

Selected values of input parameters also have a great impact when evaluating consistency in series of data. Evaluation of relative consistency of $ApEn$ in two different time-series is done by considering the variations of $ApEn$ in each data series, when the values of m and r are changed. For example, consider two time $-$ series P and Q, and two sets of inputs parameters (m_1, r_1) and (m_2, r_2). The $ApEn$ will be said to be relatively consistent if $ApEn$ (m_1, r_1) (P) $\leq ApEn$ (m_1, r_1) (Q), then $ApEn$ (m_2, r_2) (P) $\leq ApEn$ (m_2, r_2) (Q). It means that if the time series P has more regularity compared to the time series Q for one set of input parameters, it is anticipated to be the same all other sets. Then, once represented on the graph, the plots the $ApEn$ versus r by considering different data sets must not cross over one another (Richman & Moorman, 2000).

$ApEn$ helps to measure the randomness of limited and noisy data that are usually encountered in clinical time-series (Richman & Moorman, 2000). However, $ApEn$ is a biased statistic particularly for short-term data (Li et al., 2018). The bias in *ApEn* results from self-matching of each template while calculating the correlation integral $C_i^m(r)$. In other words, when computing $C_i^m(r)$, the vector $x(i)$ counts itself in order to keep the logarithms defined. If the number of matches between templates is low, the bias can reach up to 20% or 30% (Delgado-Bonal & Marshak, 2019). Excluding self-matches from the *ApEn* algorithm would be a direct means of eliminating the bias of *ApEn*. However, the absence of self-matches will leave the *ApEn* algorithm undefined unless c_i^{m+1} (r) is greater than zero for every i. *ApEn* statistics are highly sensitive to outliers, when self-matches are eliminated. Therefore, for majority of practical applications, when calculating *ApEn* statistics, self-matches cannot simply be ignored (Richman & Moorman, 2000).

5.2.3.3 Sample entropy

Sample entropy (*SampEn*) is another type of statistic to measure the randomness of a time-series. *SampEn* was developed for the purpose of solving the problems of bias and lack of relative consistency encountered in *ApEn* algorithm. Furthermore, the dependency of *ApEn* value on both length of the data series and the value of *r* can lead to a higher degree of regularity compared to the real value (Delgado-Bonal & Marshak, 2019). *SampEn* (*m, r, N*) is the negative value of the logarithm of the conditional probability that two similar sequences of *m* points remain similar at the next point $m + 1$, where self-matches are eliminated while computing condition probabilities. *SampEn* preserves the relative consistency and it does not dependent of the length of the series (Delgado-Bonal & Marshak, 2019). Unlike *ApEn* that requires a match for each template, the *SampEn* does not use a template-wise approach to determine conditional probabilities. To be defined, *SampEn* requires only that one template find a match of length $m + 1$, considering the entire series (Richman & Moorman, 2000).

Small *SampEn* values suggest regularity in patterns within the signal, whereas the opposite is an indication of greater complexity (Kapidžić et al., 2014). The *SampEn* value of a time-series of N points $\{u(i), 1 \leq i \leq N\}$, can be determined using the following algorithm (Li et al., 2016):

1. Form $(N-m\tau)$ vectors

$$x(i) = \{u(i), \ u(i+\tau), \ ..., u(i+(m-1)\tau)\}, \ 1 \leq i \leq N - m\tau \quad (5.13)$$

Here *m* is the embedding dimension whereas τ is the time delay.

2. Calculate the distance between $x(i)$ and $x(j)$ as

$$d[x(i), x(j)] = \max_{k=1,2,\dots,m} \left(|u(i+k) - u(j+k)|, 0 \leq k \leq m-1 \right).$$

$$(5.14)$$

3. Denote $A_i^{(m)}(r)$ the average number of vectors $x(j)$ within r of $x(i)$. In other words, $d\left[x(i), x(j)\right] \leq r$ for all $j = 1, 2, \dots N - m\tau$ and $j \neq i$ to avoid self-matches. In the same way, define $A_i^{(m+1)}(r)$ to determine to the similarity between vectors with next point added in the comparison.
4. Calculate the *SampEn* value of the time series $\{u(i), 1 \leq i \leq N\}$ as:

$$SampEn\,(m, \tau, r) = -\ln \frac{\sum_{i=1}^{N-m\tau} A_i^{m+1}(r)}{\sum_{i=1}^{N-m\tau} A_i^{m}(r)} \qquad (5.15)$$

SampEn provides an estimation of time-series complexity. *SampEn* is an alternative statistic to measure regularity of time-series and it can be a useful tool when one is studying the dynamics of human cardiovascular physiology (Richman & Moorman, 2000).

5.2.4 Complementary features for CO_2 signal quantification

ApEn and *SampEn* are extensively used for quantifying regularity in physiological and clinical time-series. In the respiratory signal, previous studies were limited to the assessment of irregularity in the breathing pattern of patients with panic disorder (Caldirola et al., 2004), and discriminating various types of asthma (atopic and nonatopic asthma) (Raoufy et al., 2016). Higher entropy was obtained in baseline abnormal respiratory patterns, reflecting greater irregularity and complexity in the respiratory function. However, more investigations are required to validate this study (Raoufy et al., 2016).

5.2.5 Frequency domain features

The signal's evolution over time is visualised via time domain analysis. Frequency domain analysis, on the other hand, provides information on the distribution of signal energy over a wide range of frequencies. A pair of transforms, which incorporates mutual transformation between the time domain and the frequency domain, is one of the fundamentals of signal processing. The process of converting time domain data into frequency domain can be performed using Fourier transform (FT) (Blinowska & Zygierewicz, 2011). With the inverse Fourier transform (IFT), frequency domain data are converted back to the time domain. Mathematical expressions of the FT and IFT of a time domain signal $x(t)$, are defined by Eqs. (5.16 and 5.17) (Subasi, 2019):

$$X(f) = \mathcal{F}\{x(t)\} = \int_{-\infty}^{\infty} x(t)\, e^{-j2\pi ft}\, dt \qquad (5.16)$$

$$x(t) = \mathcal{F}^{-1}\{x(f)\} = \int_{-\infty}^{\infty} X(f)\, e^{j2\pi ft}\, df \qquad (5.17)$$

where $\mathcal{F}\{.\}$ and $\mathcal{F}^{-1}\{.\}$ corresponds the operators of the FT and IFT, respectively; $x(t)$ is the time domain signal and its Fourier transform is $X_{(f)}$. The $X(f)$ function can be represented in a complex-valued form by (Eq. 5.18):

$$X(f) = |X(f)| e^{j\theta(f)} \qquad (5.18)$$

where $|X(f)|$ indicates the amplitude spectrum of the frequency components, $\theta(f)$ is the phase. Basically, the FT can be regarded as a spectral decomposition process whereby a time domain signal is represented in terms of its frequency amplitude and phase (Cassani & Falk, 2018). The discrete Fourier transform (DFT) of an N-point of the sequence $x(n)$, $n = 0, ..., N - 1$ can be defined as (Eq. 5.19) (Subasi, 2019):

$$X(k) = \sum_{n=0}^{N-1} x(n) e^{-\frac{j2\pi nk}{N}} \qquad (5.19)$$

where k shows the harmonic number. From a computational point of view, $X(k)$ must be allowed to be comprised of both positive and negative values for k. It implies that $k = -N/2..., N/2 - 1$ (Subasi, 2019). In Eq. (5.19), $x(n)$ can be either real-valued or complex. Therefore, for each value of k, the order of N complex multiplications and N complex additions is required for evaluating $X(k)$ (Kazemi et al., 2013).

Fast Fourier Transform (FFT) is a computationally appropriate algorithm for implementing the DFT and it reduces the required calculations from N^2 to $2logN$ (Subasi, 2019). FFT is substantially employed, which involves analyzing the frequency contents of a continuous time signal (Subasi, 2019). Direct interpretation of the FFT output is a complex task. Thus, the time signals are frequently characterized by their corresponding power spectrum $|X(f)|^2$ (Cassani & Falk, 2018). The power spectrum has symmetric values around the mid-point. Therefore, only half of the spectrum consisting of positive frequencies can successfully represents the signal (Van Drongelen, 2018). The spectral analysis output can also be described by the square root of the power spectrum known as amplitude spectrum. The amplitude of a time domain signal can be related to its corresponding amplitude in the frequency domain by normalizing the spectrum by $2/N$ and N is to the length of digital data (Van Drongelen, 2018). An alternative way of representing a signal in the frequency domain, is to use its phase spectrum displaying the phase versus frequency (Sicic & Boashash, 2001).

Spectral analysis is an advanced method to analyze many biomedical signals (Subasi, 2019) based on the assumption that those signals are static (Bianchi et al., 2000). The assumption may not be preserved due the variation in the underlying mechanisms of the signal generation (Chua et al., 2010). While developing the model to represent the human respiratory signal, it was thought that it was a nonstationary signal (Liu et al., 2018). Human respiration, on the other hand, is thought to be a slowly time changing process based on physiological homeostasis (Cicone & Wu, 2017). Understanding time-varying and spectral nature of a human respiratory CO_2 signal can allow to develop appropriate signal analysis methods for quantitative analysis of capnogram waveform, which can subsequentially help to identify different respiratory conditions. Autoregressive and Fast Fourier transform modeling have been applied for the estimation of the power spectral density (PSD) (Kazemi et al., 2013). The findings of this study show that asthmatic capnograms are characterized by two frequency components while capnograms recorded from subjects without asthma have one components. The first component for both asthmatic and nonasthmatic capnograms has no great difference in the magnitude (Kazemi et al., 2013).

Recently, frequency components analysis was performed on normal and asthmatic patients using FFT (Alexie & Malarvili, 2021). The preliminary results show that, the spectrum of capnogram have one main-lobe with almost equal frequencies. In addition, a few numbers of side lobes are also identified (Fig. 5.3) (Alexie & Malarvili, 2021).

Both nonasthmatic and asthmatic capnograms have one main lobe with equal frequencies. But asthmatic capnogram has higher magnitude. Table 5.2 shows the frequency components of capnogram and their corresponding magnitude for asthamatic and non-asthmatic condition.

The previous study (El-Badawy et al., 2020) has identified distinguished peaks for regular and irregular capnogram. The frequency deviation can reflect the random modification in the CO_2 level, which is identified in abnormal respiratory cycles. Capnogram pattern is affected by the condition of the gas exchange and gas flow in the lung during each breath (Krauss et al., 2005). In the subjects without abnormality in the respiratory airways,

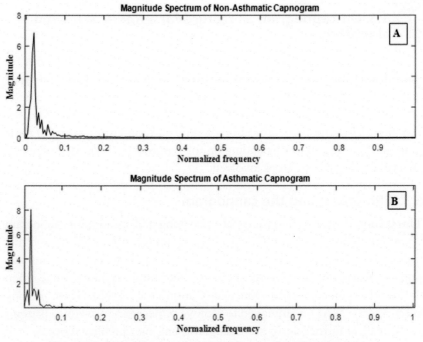

Figure 5.3 Main-lobe in asthmatic and nonasthmatic capnogram.

Table 5.2 Frequency components of capnograms and their corresponding magnitudes (Alexie and Malarvili, 2021).

	Frequency components		
	Main lobe		
Health status	**Frequency**	**Magnitude**	**Number of side lobes**
Nonasthma	0.02	6.85	3
Asthma	0.02	8.055	2

the maximum value of the exhaled CO_2 (EtCO$_2$) at the end of expiration varies between 35 and 45 mmHg (Jaffe, 2017). The patients with respiratory disorders may experience hypoventilation, a state of high partial pressure of CO_2 (PaCO$_2$). Conversely, in the patients with pulmonary edema, where the exchange of alveolar oxygen is impaired, the CO_2 level in the blood is reduced as a result of increased ventilation (Hunter et al., 2015). The capnogram peaks may be used as a complementary feature for interpreting capnogram in asthma monitoring.

5.3 On capnogram as feature for asthma detection algorithm

Over the past few years, numerous studies have been carried out for this purpose. In general, the steps involved are preprocessing of recorded capnograms, valid breath cycle identification, segmentation of capnogram, extraction of capnogram features, selection of relevant capnogram features, and then classification/identification of asthmatic conditions. Some of these steps may not be applied depending on the purpose of the study. Following are elaborations on the steps involved.

5.3.1 Preprocessing the capnogram

This section discusses the process of sharpening and smoothing capnogram waveform. The recorded CO_2 signal is contaminated by artifacts resulting from re-breathing, cardiogenic oscillation, or equipment failures. Reinspiration of exhaled breaths may occur due to a fluttering expiratory valve when the patient is mechanically ventilated (Carlon et al., 1988). Thus, moving average filter (MAF) is employed to remove unnecessary waveform patterns that can affect further signal processing. Moving average filter is simple, and it has optimal capability for a common task of reducing random noise while maintaining the shape of the signal. The smoothing action of the moving average filter decreases the amplitude of the random noise (Smith, 1997). As its name suggests, the moving average filter works in a such way that a number of points from the input signal are averaged in order to produce each point in the output signal (Guiñón et al., 2007). The difference equation of the moving average filter is expressed as Eq. (5.20) (Smith, 1997):

$$y[n] = \frac{1}{M} \sum_{j=0}^{M-1} x[n+j] \tag{5.20}$$

where $x[]$ is the input signal, $y[]$ is the output signal, M is the number of points in the average.

Another option is that a set of points of the input signal can be selected symmetrically around the output point. In that case M must be an odd number. Mathematically, this is expressed by Eq. (5.21) (Guiñón et al., 2007):

$$y[n] = \frac{1}{M} \sum_{j=\frac{-(M-1)}{2}}^{\frac{(M-1)}{2}} x[n+j] \tag{5.21}$$

The output signals obtained with M–points moving average filter can be symmetrical or directional as it is shown by Eqs. (5.20) and (5.21). Both alternatives produce a sharp noise reduction, however, the directional filter with the points on only one side, causes a relative shift between the input and output signals. The number of points on the average has the effect on the output signal. As the number of points in the filter increases, noise signal becomes lower; however, there are significant distortions of the signal by the smoothing filter (Guiñón et al., 2007). The only real parameter that can be controlled in the moving average filter is the window size, which is the amount of previous signal entries that can be averaged together. If the window is too small, the signal may still end up being very noisy and if the window is too large, critical information in the signal may be lost. The appropriate window size of the filter is selected by trial and error method (Singh, Palaniappan, et al., 2018). Window size of eight was chosen as a proper size of the moving average filter points in a study conducted for the purpose of differentiating asthmatic from non –asthmatic subjects through quantitative analysis of capnogram (Singh, Palaniappan, et al., 2018).

5.3.2 Valid breath cycles

Outlier exhalation relates the pathologic capnogram that exhibits more irregular and chaotic signal patterns compared to the healthy capnogram, which seems to have a consistent shape (Asher et al., 2014). Outlier exhalations are frequently identified in patients with complaint of asthma (Landis & Romano, 1998). Fig. 5.4A and B illustrate the CO_2 signal recorded from a healthy subject with red circles representing outlier exhalations. Validated CO_2 signal seems to have uniform breath cycles, whereas the outlier CO_2 signal is composed of disordered breath cycles (Singh, Palaniappan, et al., 2018).

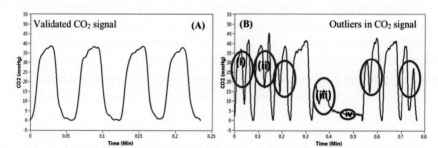

Figure 5.4 (A) validated CO_2 signal; (B) Outlier CO_2 signal (red circled CO_2 signals represent: i) rebreathing condition; ii) overriding mechanical ventilation during breathing; iii) cardiogenic oscillations; iv) breath-hold or short halt inspiration and then resumed breath) (Singh, Palaniappan, et al., 2018).

The algorithm for outlier removal was developed by overlaying exhalations to form a template exhalation view. Of these, the average exhalation was computed by taking the mean PetCO$_2$ at each time point in the superposed exhalations followed by calculating the standard deviation to identify the range of variation from the mean. Exhalation cycle was excluded if it highly deviates from a specified standard deviation (Asher et al., 2014). Valid breaths have been chosen by computing the $\left(\frac{dCO_2}{dt}\right)$ and the mean of each breath phase. The positive mean was assumed to reflect exhalation phase, whereas negative mean corresponded to the inspiration phase (Singh, Palaniappan, et al., 2018).

5.3.3 Segmentation of capnogram

After removing outliers, selected breaths are subject to the segmentation process in order to identify components of interest for further signal analysis. Segmentation of a capnogram waveform into small parts from the inspiratory and expiratory phases, facilitates the interpretation and better understanding of the signal (Bhavani-Shankar & Philip, 2000). Moreover, it helps to localize features in different segments containing only relevant information instead of considering the whole signal including portions with no significant information for the signal analysis (Betancourt et al., 2014).

In different studies, segmentation was performed manually after setting the time-based criteria for selecting appropriate breath cycles. In a study of You et al. (1994), the minimum and maximum duration of expiratory phase were 0.8 and 3 s, respectively. Later, Howe, Teo Aik et al. computed the slope of phase II lasting only for 0.25 s from the onset of upstroke phase and the slope of phase III as a portion of alveolar plateau lasting only for 0.5 s. The respiratory baseline was chosen to be 4 mmHg, based on the assumption that the carbon dioxide has to have come from the lungs to reach a level of 4 mmHg (Howe et al., 2011). The duration of the expiratory phase greatly affects alveoli plateau phase of the waveform. In some circumstances, expiration lasting for less than 1.5 s or greater than 2.5 s, may result to the under-estimation or over-estimation of bronchospasm (You et al., 1994). A threshold method was applied to segment capnogram waveform by considering the expiratory CO$_2$ (mmHg) on the y-axis instead of the exhalation duration on the x-axis (time) (Singh, Palaniappan, et al., 2018). The respiratory baseline of each breath cycle was chosen to be 4 mmHg by referring to the previous studies (Howe et al., 2011; You et al., 1994). Fig. 5.5 shows different segments of a complete breath cycle obtained by limiting each phase between a certain range of the CO$_2$ output using the threshold approach.

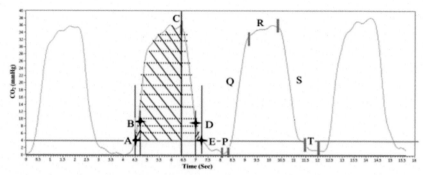

Figure 5.5 Extracted areas in recorded capnogram waveform are marked with alphabets (Singh, Palaniappan, et al., 2018).

The upward expiratory phase (A–B) lasted between 4 and 10 mmHg, then the absolute expiratory phase (A–B–C) was restricted between 4 mmHg (i.e., at the onset of exhalation) to the peak value of expiratory CO_2. Thereafter, the downward phase (D–E) was limited between 10 and 4 mmHg on the descending part of the breath cycle. Segmented parts were then considered while extracting features for identifying asthmatic conditions (Singh, Palaniappan, et al., 2018).

5.3.4 Feature extraction of capnogram

Feature extraction has the impact on achieving the optimal performance in classifying capnogram features. Feature extraction deals with finding features from the original data set, which are most informative and nonredundant to enhance the effectiveness of further signal processing and analysis (Ramachandran et al., 2020). Although many attempts have been made for extracting features from capnogram, most of them used manual technique (Yaron et al., 1996; You et al., 1994). In fact, manual analysis of the signal is time consuming and there is a possibility of producing errors. Linear trendline method was applied to compute slopes of the ascending phase and alveolar plateau phase and the intermediate angle (alpha angle) was derived (Howe et al., 2011). Linear trendline is mathematically defined by (Eq. 5.22):

$$y = mx + c \tag{5.22}$$

where m corresponds to the slope of the segment analyzed. An obtuse (α) angle is formed by the slope of the ascending phase and plateau line.

Trendline analysis requires the data to be distributed linearly, which is utterly laborious in real time because of the uneven removal of CO_2 samples from the alveoli. Thus, capnogram waveform was considered to resemble a series of quadratic parabolic segments rather than straight lines. Afterward, Simpson's rule was applied to compute different areas (AR_i) of each quantized respiratory cycle. Those areas were used as time domain features to characterize asthmatic changes (Singh, Palaniappan, et al., 2018). The areas (AR_i) were calculated using Eq. (5.23):

$$A_m = \frac{dt}{6} \sum_{n=0}^{m} (C_{n-1}(t) + 4C_n(t) + C_{n+1}(t)) \qquad (5.23)$$

where $C(t)$ and $d(t)$ represents the CO_2 signal and the sampling interval, respectively. All the methods mentioned above, have been applied for the extraction of time or frequency domain features. However, some useful information might be disregarded when the signal is only analyzed in the time or in the frequency domain. In fact, time domain analysis does not provide spectral information of the signal. Similarly, frequency domain lacks the representation of time related information about the signal (Betancourt et al., 2014). Representation of the signal simultaneously in both time and frequency domain can improve the analysis of capnogram waveform. Betancourt et al. extracted wavelet coefficients from capnogram segments for classifying asthma severity levels. Breath cycles were first decomposed into small segments (A-B, C-D, E-F and G-H) (Fig. 5.6) (Betancourt et al., 2014).

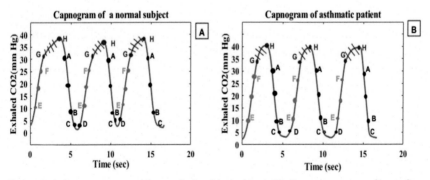

Figure 5.6 Capnogram waveform of a healthy subject (A); Capnogram waveform of an asthmatic patient (B); (F—E), (G—H) and (A—B) represent segments from the upstroke expiratory, alveolar plateau, and downward inspiratory phases, respectively (Betancourt et al., 2014).

5.3.5 Identification of asthmatic/nonasthmatic conditions

Capnogram signal analysis is critical as it determines the variations in capno-gram waveform and consequently differentiates the airway illness. Presented is a method for differentiating asthma and nonasthmatic conditions. The CO_2 waveform was recorded from each patient during pre- and posttreat-ment and processed for individuals. Firstly, the segments were divided each four valid breath cycle into sub-cycles by employing simple threshold method, on the contrary to manual and visual inspection. Each breath cycle of four breath cycles of each patient was segmented into two regions 4−10 mmHg, and 11−15 mmHg, by creating threshold as presented in Fig. 5.7. In addition, an alveolar phase was parted from each breath cycle for 0.75 s by recording for 1 s from End-tidal Point and subtracting 0.25 s to ensure the constancy of points of measurement.

Further, two features, *area* (AR_i) and *slope*, were computed from the segmented part of each breath cycle using Eqs. (5.24) and (5.25). The slope

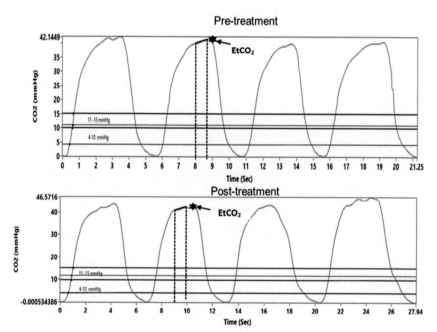

Figure 5.7 Illustration of segmented part of each breath cycle from four breath cycles. The first region enclosed with lines reflects the lower part of the upward expiratory phase (4−10 mmHg), next enclosed region indicates the upper part of the upward expiratory phase (11−15 mmHg), the enclosed area in alveolar phase depicts the 0.75 s recording from the 0.25 s of end-tidal CO_2, and the symbol star represents the end-tidal CO_2 point.

of the upward expiratory phase (i.e., 4—10 mmHg, 11—15 mmHg) and alveolar phase was estimated using the general least squares linear fitting method. It computes the intercept and slope of the CO_2 waveform by lessening the residue according to the Eq. (5.25), which possibly permit the inclusion of all-out CO_2 signal.

$$AR_i = \frac{dt}{6} \sum_{j=0}^{i} \left(R_{j-1}(t) + 4R_j(t) + R_{j+1}(t) \right) \tag{5.24}$$

where, dt and $R(t)$ signify the sampling interval and CO_2 signal, respectively.

$$Slope(S) = \frac{1}{C} \sum_{j=0}^{C-1} b_j \left(M_j - S_j \right)^2 \tag{5.25}$$

where C is the length of slope (S), which reflects the CO_2 signal, b_j, and M_j are the jth element of Weight and best linear fit, respectively and S_j, is the jth element of S. The CO_2 signal was recorded from the asthmatic patients after the diagnosis, just before the onset of treatment. The posttreatment CO_2 data was recorded after the completion of treatment and was considered fit for release.

The CO_2 waveform, measured via a capnography, indicates variations in bronchospasm that reflect the heterogeneity of the exhaled air. A decrease and increase in *area* and *slope,* respectively, for the lower part of the expiratory (4—10 mmHg) and alveolar phase were observed. Where the change in the morphology of CO_2 waveform can be seen from the bare view, for example, "shark fin" look, more subtle variations require computation of other features. As this is a new concept, most of the studies to date, this has been performed manually (You et al., 1994; Yaron et al., 1996). Presented study shows a slightly different method. Keeping in mind that this concept may be possibly implemented a lately developing human respiration CO_2 measurement device (Singh, Howe, et al., 2018), a simple threshold method was employed for the segmentation of complete breath cycle into sub-cycle. For the *slope* and *area,* the first part was restricted between 4 and 10 mmHg because it is believed that CO_2 must come from lungs to reach to 4 mmHg, second part was limited between 11 and 15 mmHg, whereas third region, part of alveolar phase was confined for 0.75 s as recorded for 1s from end-tidal point and subtracted for 0.25 s. These indices (AR_i and *slope*) were extracted using Simpson's rule and general least-squares linear fitting method.

The mean AR_i and Slope for the pretreatment were significantly higher and steeper, respectively than the posttreatment for both upward expiratory and alveolar phase. The mean value of AR_i of the asthmatic patient for the lower part (4−10 mmHg) of the upward expiratory phase was higher than the medicated patients, and the slope was steeper with the medicated patient. Besides, AR_i for the upper part (11−15 mmHg) of the expiratory phase was more moderate before the medication compared with a medicated patient, whereas slope was minimum with asthmatic before treatment compared to that of the after -treatment medication. Besides, AR_i of alveolar phase was lower before treatment than that of the after-treatment, with minimum deviation from its mean. However, the difference of mean and SD was significantly higher (mean, 4.16) and lower (SD, 0.28) respectively, for the alveolar phase compared with upward expiratory phases during pre- and posttreatment. Besides, the pretreatment mean slope values were markedly steeper compared with posttreatment. Moreover, the mean and SD differences were 0.08 and 0.06 for the upward expiratory phase (4−10 mmHg) than that of the other parts. Table 5.3 shows the parameters computed from capnogram waveform, before and after treating the patients.

Table 5.3 Parameters computed from capnogram waveform, before and after treating the patients.

Pretreatment

Segmented part	Features	Cycle 1	Cycle 2	Cycle 3	Cycle 4	Mean
4−10 mmHg	Area	1.10	1.02	1.17	1.29	1.17
	Slope	0.36	0.36	0.34	0.31	0.34
11−15 mmHg	Area	1.26	1.37	1.38	1.50	1.38
	Slope	0.46	0.44	0.41	0.38	0.42
Alveolar Phase	Area	30.63	29.78	28.32	29.47	29.55
	Slope	0.01	0.02	0.04	0.01	0.02

Posttreatment

Segmented part	Features	Cycle 1	Cycle 2	Cycle 3	Cycle 4	Mean
4−10 mmHg	Area	1.35	1.24	1.03	1.30	1.27
	Slope	0.29	0.33	0.38	0.30	0.31
11−15 mmHg	Area	1.62	1.77	1.25	1.49	1.54
	Slope	0.36	0.34	0.43	0.38	0.38
Alveolar Phase	Area	31.20	30.61	30.93	33.71	31.61
	Slope	0.01	0.02	0.03	0.003	0.01

Besides, the lower part (4–10 mmHg) of the upward expiratory and alveolar phase changes were significantly dissimilar pre- and posttreatment. In other words, the indices extracted from the CO_2 waveform were able to detect improvement after posttreatment in the patient suffering from bronchospasm, as could PFM and clinical parameters. Respired CO_2 measurement has some advantages. The study verified that a change in the morphology of CO_2 waveform represents a change in airway obstruction. Respired CO_2 measurement device has the additional benefit of being patient independent, by this means reducing the influence of patient's cooperation and understanding on monitoring parameters. On the other hand, although the specific CO_2 morphology changes in asthma was observed, these required further verification on large numbers of the subjects to confirm the certainty of CO_2 waveform indices, which are reflective of asthmatic conditions. Nevertheless, it is justified that the measurement of CO_2 waveform has excellent potential for the monitoring of asthma severity due to its unique properties of continuous tracking, noninvasive, and effort independent.

Furthermore, the added advantages of this mechanism are that the monitoring device does not interfere with the ease of initiating therapy nor require active patient efforts. Further, the work now lies in recognition of proper waveform, implementation of the proposed indices into the real-time CO_2 measurement device, and the reporting of such indices in reproducible and an easily understood form that would significantly enhance the field of asthma monitoring. It is believed that this will assist a better sympathetic of asthma, improved management, and eventually a decrease in morbidity and mortality.

References

Alexie, M., & Malarvili, M. (2021). Investigation on properties of capnogram: On stationarity and spectral components of the signal. In *2021 IEEE National Biomedical Engineering Conference (NBEC)*.

Asher, R.J., Heldt, T., Krauss, B.S., & Verghese, G.C. (2014). *U.S. Patent Application No. 13/849,284*.

Betancourt, J. P., Tangel, M. L., Yan, F., Diaz, M. O., Otaño, A. E. P., Dong, F., & Hirota, K. (2014). Segmented wavelet decomposition for capnogram feature extraction in asthma classification. *Journal of Advanced Computational Intelligence and Intelligent Informatics, 18*(4), 480–488.

Bhavani-Shankar, K., & Philip, J. H. (2000). Defining segments and phases of a time capnogram. *Anesthesia and Analgesia, 91*(4), 973–977.

Bianchi, A. M., Mainardi, L. T., & Cerutti, S. (2000). Time-frequency analysis of biomedical signals. *Transactions of the Institute of Measurement and Control, 22*(3), 215–230.

Blinowska, K. J., & Zygierewicz, J. (2011). *Practical biomedical signal analysis using MATLAB®*. CRC Press.

Böhm, S., Kremeier, P., Tusman, G., Reuter, D., & Pulletz, S. (2020). Volumetric capnography for analysis and optimization of ventilation and gas exchange. In *Der Anaesthesist*.

Borowska, M. (2015). Entropy-based algorithms in the analysis of biomedical signals. *Studies in Logic, Grammar and Rhetoric, 43*(1), 21−32.

Caldirola, D., Bellodi, L., Caumo, A., Migliarese, G., & Perna, G. (2004). Approximate entropy of respiratory patterns in panic disorder. *American Journal of Psychiatry, 161*(1), 79−87.

Carlon, G. C., Ray, C., Jr., Miodownik, S., Kopec, I., & Groeger, J. S. (1988). Capnography in mechanically ventilated patients. *Critical Care Medicine, 16*(5), 550−556.

Cassani, R., & Falk, T. H. (2018). Spectrotemporal modeling of biomedical signals: Theoretical foundation and applications. In *Reference module in biomedical sciences*. Elsevier.

Chua, K. C., Chandran, V., Acharya, U. R., & Lim, C. M. (2010). Application of higher order statistics/spectra in biomedical signals—A review. *Medical Engineering & Physics, 32*(7), 679−689.

Cicone, A., & Wu, H.-T. (2017). How nonlinear-type time-frequency analysis can help in sensing instantaneous heart rate and instantaneous respiratory rate from photoplethysmography in a reliable way. *Frontiers in Physiology, 8*, 701.

Cocconcelli, M., Strozzi, M., Molano, J. C. C., & Rubini, R. (2022). Detectivity: A combination of Hjorth's parameters for condition monitoring of ball bearings. *Mechanical Systems and Signal Processing, 164*, 108247.

Delgado-Bonal, A., & Marshak, A. (2019). Approximate entropy and sample entropy: A comprehensive tutorial. *Entropy, 21*(6), 541.

El-Badawy, I. M., Omar, Z., & Singh, O. P. (2022). *An effective machine learning approach for classifying artefact-free and distorted capnogram segments using simple time-domain features*. IEEE Access.

El-Badawy, I. M., Singh, O. P., & Omar, Z. (2020). Automatic classification of regular and irregular capnogram segments using time-and frequency-domain features: A machine learning-based approach. *Technology and Health Care (Preprint)*, 1−14.

Ferlazzo, E., Mammone, N., Cianci, V., Gasparini, S., Gambardella, A., Labate, A., Latella, M. A., Sofia, V., Elia, M., & Morabito, F. C. (2014). Permutation entropy of scalp EEG: A tool to investigate epilepsies: Suggestions from absence epilepsies. *Clinical Neurophysiology, 125*(1), 13−20.

Glenny, R. W. (2011). Emergence of matched airway and vascular trees from fractal rules. *Journal of Applied Physiology, 110*(4), 1119−1129.

Gravenstein, J. S., Jaffe, M. B., Gravenstein, N., & Paulus, D. A. (2011). *Capnography*. Cambridge University Press.

Guiñón, J. L., Ortega, E., García-Antón, J., & Pérez-Herranz, V. (2007). Moving average and Savitzki-Golay smoothing filters using Mathcad. *Papers ICEE*, 1−4.

Guthrie, B. D., Adler, M. D., & Powell, E. C. (2007). End-tidal carbon dioxide measurements in children with acute asthma. *Academic Emergency Medicine, 14*(12), 1135−1140.

Hadoush, H., Alafeef, M., & Abdulhay, E. (2019). Brain complexity in children with mild and severe autism spectrum disorders: Analysis of multiscale entropy in EEG. *Brain Topography, 32*(5), 914−921.

Hisamuddin, N. N., Rashidi, A., Chew, K., Kamaruddin, J., Idzwan, Z., & Teo, A. (2009). Correlations between capnographic waveforms and peak flow meter measurement in emergency department management of asthma. *International Journal of Emergency Medicine, 2*(2), 83−89.

Hjorth, B. (1970). EEG analysis based on time domain properties. *Electroencephalography and Clinical Neurophysiology, 29*(3), 306−310.

Hong, C. S., Ghani, A. S. A., & Khairuddin, I. M. (2018). Development of an Electronic Kit for detecting asthma in Human Respiratory System. In (1st), *319*. IOP Conference Series Materials Science and Engineering. IOP Publishing.

Howe, T. A., Jaalam, K., Ahmad, R., Sheng, C. K., & Ab Rahman, N. H. N. (2011). The use of end-tidal capnography to monitor non-intubated patients presenting with acute exacerbation of asthma in the emergency department. *The Journal of Emergency Medicine, 41*(6), 581–589.

Hunter, C. L., Silvestri, S., Ralls, G., & Papa, L. (2015). Prehospital end-tidal carbon dioxide differentiates between cardiac and obstructive causes of dyspnoea. *Emergency Medicine Journal, 32*(6), 453–456.

Jaffe, M. B. (2017). Using the features of the time and volumetric capnogram for classification and prediction. *Journal of Clinical Monitoring and Computing, 31*(1), 19–41.

Kapidžić, A., Platiša, M. M., Bojić, T., & Kalauzi, A. (2014). Nonlinear properties of cardiac rhythm and respiratory signal under paced breathing in young and middle-aged healthy subjects. *Medical Engineering & Physics, 36*(12), 1577–1584.

Kazemi, M., Krishnan, M. B., & Howe, T. A. (2013). Frequency analysis of capnogram signals to differentiate asthmatic and non-asthmatic conditions using radial basis function neural networks. *Iranian Journal of Allergy, Asthma and Immunology*, 236–246.

Kean, T. T., & Malarvili, M. B. (2009). Analysis of capnography for asthmatic patient. In 2009 IEEE International Conference on Signal and Image Processing Applications (pp. 464–467). IEEE.

Kelsey, J., Oldham, E., & Horvath, S. (1962). Expiratory carbon dioxide concentration curve a test of pulmonary function. *Diseases of the Chest, 41*(5), 498–503.

Krauss, B., Deykin, A., Lam, A., Ryoo, J. J., Hampton, D. R., Schmitt, P. W., & Falk, J. L. (2005). Capnogram shape in obstructive lung disease. *Anesthesia & Analgesia, 100*(3), 884–888.

Kunkov, S., Pinedo, V., Silver, E. J., & Crain, E. F. (2005). Predicting the need for hospitalization in acute childhood asthma using end-tidal capnography. *Pediatric Emergency Care, 21*(9), 574–577.

Landis, B., & Romano, P. M. (1998). A scoring system for capnogram biofeedback: Preliminary findings. *Applied Psychophysiology and Biofeedback, 23*(2), 75–91.

Langhan, M. L., Zonfrillo, M. R., & Spiro, D. M. (2008). Quantitative end-tidal carbon dioxide in acute exacerbations of asthma. *The Journal of pediatrics, 152*(6), 829–832.

Li, P., Karmakar, C., Yan, C., Palaniswami, M., & Liu, C. (2016). Classification of 5-S epileptic EEG recordings using distribution entropy and sample entropy. *Frontiers in Physiology, 7*, 136.

Li, P., Karmakar, C., Yearwood, J., Venkatesh, S., Palaniswami, M., & Liu, C. (2018). Detection of epileptic seizure based on entropy analysis of short-term EEG. *PLoS One, 13*(3), e0193691.

Liu, M., Xue, H. J., Liang, F., Lv, H., Li, Z., Qi, F. G., Zhang, Z., & Wang, J. (2018). UWB-Radar-Sensed human respiratory signal modeling based on the morphological method. *Progress In Electromagnetics Research C, 88*, 235–249.

Malik, S. A., Singh, O. P., Nurifhan, A., & Malarvili, M. (2016). Portable respiratory CO_2 monitoring device for early screening of asthma. *Proceedings ACEC*.

Meyer, M., Mohr, M., Schulz, H., & Piiper, J. (1990). Sloping alveolar plateaus of CO_2, O_2 and intravenously infused C_2H_2 and $CHClF_2$ in the dog. *Respiration physiology, 81*(2), 137–151.

Mieloszyk, R. J., Verghese, G. C., Deitch, K., Cooney, B., Khalid, A., Mirre-González, M. A., Heldt, T., & Krauss, B. S. (2014). Automated quantitative analysis of capnogram shape for COPD–normal and COPD–CHF classification. *IEEE Transactions on Biomedical Engineering, 61*(12), 2882–2890.

Oh, S.-H., Lee, Y.-R., & Kim, H.-N. (2014). A novel EEG feature extraction method using Hjorth parameter. *International Journal of Electronics and Electrical Engineering, 2*(2), 106—110.

Pincus, S. M. (1991). Approximate entropy as a measure of system complexity. *Proceedings of the National Academy of Sciences, 88*(6), 2297—2301.

Pincus, S. M., & Goldberger, A. L. (1994). Physiological time-series analysis: What does regularity quantify? *American Journal of Physiology-Heart and Circulatory Physiology, 266*(4), H1643—H1656.

Ramachandran, D., Thangapandian, V. P., & Rajaguru, H. (2020). Computerized approach for cardiovascular risk level detection using photoplethysmography signals. *Measurement, 150,* 107048.

Raoufy, M. R., Ghafari, T., Darooei, R., Nazari, M., Mahdaviani, S. A., Eslaminejad, A. R., Almasnia, M., Gharibzadeh, S., Mani, A. R., & Hajizadeh, S. (2016). Classification of asthma based on nonlinear analysis of breathing pattern. *PLoS One, 11*(1), e0147976.

Richman, J. S., & Moorman, J. R. (2000). Physiological time-series analysis using approximate entropy and sample entropy. *American Journal of Physiology-Heart and Circulatory Physiology, 278*(6), H2039—H2049.

Sicic, V., & Boashash, B. (2001). Parameter selection for optimising time-frequency distributions and measurements of time-frequency characteristics of non-stationary signals. In *2001 IEEE International Conference on Acoustics, Speech, and signal processing. Proceedings (Cat. No. 01CH37221).*

Singh, O. P., Howe, T. A., & Malarvili, M. (2018). Real-time human respiration carbon dioxide measurement device for cardiorespiratory assessment. *Journal of Breath Research, 12*(2), 026003.

Singh, O. P., Palaniappan, R., & Malarvili, M. (2018). Automatic quantitative analysis of human respired carbon dioxide waveform for asthma and non-asthma classification using support vector machine. *IEEE Access, 6,* 55245—55256.

Smith, S. W. (1997). *The scientist and engineer's guide to digital signal processing.*

Subasi, A. (2019). *Practical guide for biomedical signals analysis using machine learning techniques: A MATLAB based approach.* Academic Press.

Suki, B. (2002). Fluctuations and power laws in pulmonary physiology. *American Journal of Respiratory and Critical Care Medicine, 166*(2), 133—137.

Thamrin, C., Frey, U., Kaminsky, D. A., Reddel, H. K., Seely, A. J., Suki, B., & Sterk, P. J. (2016). Systems biology and clinical practice in respiratory medicine. The twain shall meet. *American Journal of Respiratory and Critical Care Medicine, 194*(9), 1053—1061.

Tolnai, J., Fodor, G. H., Babik, B., Dos Santos Rocha, A., Bayat, S., Peták, F., & Habre, W. (2018). Volumetric but not time capnography detects ventilation/perfusion mismatch in injured rabbit lung. *Frontiers in Physiology, 9,* 1805.

Van Drongelen, W. (2018). *Signal processing for neuroscientists.* Academic Press.

Van Meerten, R. (1971). Expiratory gas concentration curves for examination of uneven distribution of ventilation and perfusion in the lung. *Respiration, 28*(2), 167—185.

Vargha, G. (1976). Evaluation of some simple methods of expressing the capnographic curve. *Pneumonologie, 153*(2), 105—108.

Xiang, J., Li, C., Li, H., Cao, R., Wang, B., Han, X., & Chen, J. (2015). The detection of epileptic seizure signals based on fuzzy entropy. *Journal of Neuroscience Methods, 243,* 18—25.

Yaron, M., Padyk, P., Hutsinpiller, M., & Cairns, C. B. (1996). Utility of the expiratory capnogram in the assessment of bronchospasm. *Annals of Emergency Medicine, 28*(4), 403—407.

You, B., Peslin, R., Duvivier, C., Vu, V. D., & Grilliat, J. (1994). Expiratory capnography in asthma: Evaluation of various shape indices. *European Respiratory Journal, 7*(2), 318—323.

CHAPTER SIX

Design of carbon dioxide sensor for capnography

As discussed in previous chapters, carbon dioxide (CO_2) gas released by human body is a notable excretory product from cellular respiration of living organism in identifying several respiratory and pulmonary illness. Among many, asthma as the primary respiratory diseases is one of the key diseases that can be diagnosed with the evaluation of parameters identified from CO_2 released by a patient. Regarding that, Chapter 3—5 have detailed about capnography, a noninvasive device that uses sensing technology and measures human respiration CO_2 from the expired gasses and an incessant plot of exhaled CO_2 over time, known as a capnogram. Capnography detects CO_2 from expired gas and extracts the featured CO_2 parameters such as end-tidal CO_2 (EtCO2), respiratory rate (RR), slope angles and many more, which can be used to differentiate asthmatic and nonasthmatic conditions. For an effective capnography, a high-performance CO_2 sensor is necessary. There are a variety of CO_2 sensors developed and a huge effort is involved in improving the technology for highly sensitive CO_2 sensors. The present chapter discusses the types of CO_2 sensors and its contribution for diagnosing asthma, as well as in enhancing capnography.

6.1 Evolution of carbon dioxide sensing

The composition of on Earth varies substantially and continuously since the planet formed 4500 million years ago. The substantial changes are due to the variation in terrestrial atmosphere, which always fluctuates the composition of gas. However, the variation remains in a certain threshold as it does not affect the living and geographical cycles. The minor fluctuations impact the evolution of land and aquatic organisms and to sculpt the geographical elements over the years (Dervieux et al., 2021; Tang et al., 2011). CO_2 is the most influential gas in the atmosphere after oxygen (O_2), which varies with geographical changes. Geologists have reported that CO_2 gas fluctuated at 7000 ppm in the atmosphere at the time of Cambrian period. At present condition, CO_2 gas fluctuates at 400 ppm at maximum pressure, which influenced the current nature of life on earth. At

Systems and Signal Processing of Capnography as a Diagnostic Tool for Asthma Assessment
ISBN: 978-0-323-85747-5
https://doi.org/10.1016/B978-0-323-85747-5.00004-8

initial days, CO_2 gas was monitored for climate prediction and as precaution procedure for natural disasters. Over the time, the measurement of CO_2 in atmosphere was given equal important for determining the condition and niche of microenvironments, involving the living organism and decomposing matter. The study correlates with every element and aspect of individual field as the respiring organism greatly influences the CO_2 level in the atmospheric layer (Blombach & Takors, 2015; Cummins et al., 2014). Being the only source of CO_2 production, living organism produce CO_2 in the citric cycle as the byproduct of respiration metabolism. CO_2 is in high concentration in the living organism compared to external atmosphere. The differences of CO_2 gradient in and out of living organism develop several mechanisms to detect the CO_2 variation and its influences in various fields. Living organisms consist of bacteria, fungi, plants, insects, animals, and humans portray a significant role in the changes and impacts of CO_2 variation in food, environment, and medical applications (Kroukamp & Wolfaardt, 2009; Neethirajan et al., 2010). CO_2 gas is used as indicator in food quality determination to detect the microbial growth rate and subsequently the food spoilage period, mainly in food packaging. CO_2 alone or integrated with other gases develops a protective layer for the packaged food to eliminate O_2 entering the food. Being the major cause of food spoiling, O_2 gas is extremely inhibited in the packaged food and the level of CO_2 is observed to a certain extent for food preservations (Neethirajan et al., 2009; Puligundla et al., 2012). CO_2 is known as active packing gas in food packing, as it can lower the rate of microbial growth in food by suppressing the increasing O_2 gas. Hence, the quality and safety of food is justified by determining the concentration of CO_2 in food packing. Thereafter, detecting techniques for measuring CO_2 in food packing were developed (Chaix et al., 2014; Meng et al., 2014). CO_2 gas is one of the most important parameters measured in environmental monitoring. It is equally significant in identifying the quality of air in land and in aquaculture (Chang et al., 2021; Mendes et al., 2019). In aquaculture, CO_2 gas is determined based on the ocean acidification in term of dissolved CO_2 and atmospheric CO_2 levels. It is significant to ensure highly sensitive aqua species to be protected and good quality of water is preserved. Variation in CO_2 level changes the physical and chemical equilibrium of aqua environment, which also plays significant role for quantitative identification of CO_2 level (Berman et al., 2012; Brown et al., 2020). In addition, CO_2 sensing is one of accurate quantification in determining greenhouse gas effects. CO_2 is a fundamental component of the atmospheric carbon cycle, and it has been identified as the primary cause of climate change. Scientists need to precisely characterize the regional and temporal distribution of CO_2 to better comprehend the global carbon cycle and anticipate future climate change. Regarding environmental safety and health, CO_2 is monitored in heavy industries and large-scale manufacturing plants to prevent greenhouse effect and gas pollution (Honeycutt et al., 2021; Quan et al., 2011). In medical applications, level of CO_2 is measured as a vital sign to determine rudimentary health state of a human. Accurate measurement of CO_2 helps to identify appropriate treatments and effectual care to patients. The primary measurement of CO_2 in human through invasive blood sampling provided respiratory and circulatory indications on the health state of a patient (Cummins et al., 2020; Decker et al., 2019). With the advancement of healthcare system, noninvasive CO_2 measurement is prominent at medical facilities as the preliminary health check for a patient. As discussed in the previous chapters, there are several sophisticated medical devices to measure CO_2 and classify the measurement into partial pressure of CO_2, end tidal CO_2, respiratory rate and many more digital data to justify the respiratory condition of patient (Liu et al., 2022; Siefker et al., 2021). Being the very critical element in CO_2 measurement in medical devices, CO_2 sensors are prominently developed for high precision measurement and applications. As the CO_2 measurement is vital in food and environmental condition, the CO_2 sensors have gone through a series of development with several measurement techniques, which suites the application (Mills, 2009). However, the advancement at each stage has impacted the state-of-the-art of CO_2 sensors in medical applications. This chapter discusses the existing sensor techniques for CO_2 measurement in medical applications, especially in asthma monitoring.

6.2 Carbon dioxide sensors

Enhanced carbon dioxide (CO_2) detection with exceptional sensitivity and rapid reaction remains crucial in the medical area, particularly in the diagnosis of asthmatic condition. A wide range of sensors has been introduced to measure CO_2 from human breath. Nonetheless, the efforts to develop better CO_2 sensing tools are significant with the enhancement of global technology (Ghorbani & Schmidt, 2017). Exhaled CO_2 measurement is significant for noninvasive diagnosis of asthma monitoring by assessing CO_2 partial pressure (pCO_2) and end-tidal CO_2 ($EtCO_2$). Noninvasive CO_2 measurement provides outputs to medical specialists about a patient's asthmatic status and for further evaluations by monitoring breathing, perfusion, and systemic metabolism. The noninvasive CO_2 measurement from human breath is an emerging approach of interest for examining the metabolic state of human asthmatic condition and determining airway blockage and disease. As a result, high–performance analytical equipment for real-time CO_2 measurement for monitoring asthma is in high demand. The influence from humidity in breath samples is a common problem with noninvasive CO_2 measurement tools. In the literature, several sample preparation procedures for removing breath humidity have been discussed. Unfortunately, it raises the expense of product technology and limits the use of measuring tools outside of clinical settings (Zhao et al., 2014). In this regard, research with alternative CO_2 sensing has been developed by introducing a variety of CO_2 sensing tools with varying methods of CO_2 collection, analysis, and output presentation, including relatively simple, inexpensive, rapid, and user-friendly tools in clinical settings and at home environments. The chapter provides an overview of the different types of CO_2 sensors that have been created for using human breath to monitor asthma.

6.3 Optical CO_2 sensors

In the expanding demand for monitoring asthma and other healthcare illness, optical sensors hold higher rate than other sensors. The sensors are widely used as it suits the needs of current electronic industries with low-cost, and real-time technology. Optical sensors were first presented in the late 1970s, when the mechanism of light-based sensing was discovered. The sensors are used in fiber technology, which fine-tunes light characteristics to produce a desired transducer in a sensing system (Fritzsche et al., 2017; Yoo et al., 2010). When a phenomenal change happens in a system, an optical sensor is modulated with the optical properties of lights, such as wavelength, intensity, polarization, distribution, and many others to interpret and resolve the issue (Fajkus et al., 2017; Ogawa et al., 2018). With the use of dielectric materials and components, optical sensors have a high level of sensitivity and adaptability. The sensors have good compatibility and immunity to electromagnetic interferences due to its chemically inert property. As a result, optical sensors are adaptable to harsh environments, such as the interior of an electronic construction block or a transmission system, where conventional sensors would fail. Another optical approach for sensing CO_2 is fiber optics. Optodes are the name for these types of sensors. At the tip of a fiber optic sensor is a chemical sensing layer that alters optical characteristics in response to CO_2. The principle of optical-fiber-based CO_2 sensors relies on the difference in absorbance or reflectance upon the response of CO_2 gas. Fiber optic gas sensors are chemically inert and do not cross-react with the gas in the background. In addition, a fiber optic gas sensor has the disadvantage that it requires frequent cleaning of the lenses to ensure continuous operation. Since fiber optic gas sensors use lenses, dust and soot can interfere with the transmission of light over time. Detecting changes at the tip of the fiber requires expensive readout equipment (Chien et al., 2020).

6.4 Electrochemical CO_2 sensors

Electrochemical sensors, known as solid-state electrolyte sensors, are the well-recognized sensors for CO_2 monitoring. Electrochemical sensors for gas detection have a high level of reliability since they can work at a wide range of temperatures (-30 to $1600°C$) for long periods of time without requiring additional heating. Electrochemical gasoline sensors have utilized as a promising biosensor to reap facts in disruptive and complex bio-environments (Aroutiounian, 2020). The flexibility of the use of more than one functional substance improves the sensor flexibility to amend in line with the desired applications (Hanafi et al., 2019; Khan et al., 2019). The sensors are classified into amperometric, conductometric, and potentiometric sensors. In the potentiometric mode, the signal is an electromotive force, whereas in the amperometric mode, an electric current is measured. Conductometric sensors analyze the current-voltage plot. The materials used to construct sensors should be able to withstand extreme temperatures and corrosive media. Solid state sensors are the best candidate for the development of commercial gas sensors due to the variety of principles and materials (metal oxides, polymers, ceramics, or sol-gel). Due to their small size, solid state electrolyte sensors able to detect very low concentrations (parts per million or even parts per billion), are available, and are practical for large-scale production, allowing them to be very affordable. However, they have limited measurement accuracy and lack long-term stability (Khan et al., 2019).

6.5 Metal oxide CO_2 sensors

A metal oxide gas sensor consists of a sensitive layer (sensing material) deposited over an electrode–equipped substrate. The sensor element is heated by a unique heater separated from the sensor layer and electrodes by an insulating layer. This heating element can be a thin layer of platinum or platinum alloy wire, resistant metal oxides, or deposited platinum. Electrodes are connected to the sensor material to form a closed circuit (Vajhadin et al., 2021). The mechanism of gas detection by the metal oxide sensor is based on the chemical reaction that occurs on the surface of the sensor. In the presence of CO_2 gas, the metal oxide dissociates the gas into charged ions or complexes, which lead to the movement of electrons. This changes the conductivity or resistance of the surface layer when exposed to the analyte. This change in conductivity is directly related to the amount of CO_2 present in the environment, leading to the identification and quantification of gas concentration. The built-in heater that heats the metal oxide material to the optimum operating temperature range for the detected gas is controlled (Oprea et al., 2018). A pair of bias electrodes are embedded in the metal oxide to measure changes in conductivity or resistance. Fig. 6.1 shows schematic of metal oxide–based gas sensor, consist of (microelectromechanical systems) MEMS substrate, heating membrane and a pair of electrodes. Sensors typically produce very strong signals at high gas concentrations. When designing a MEMS-based microelectronics sensor, the choice of sensor material is an important criterion (Vijayakumari et al., 2021).

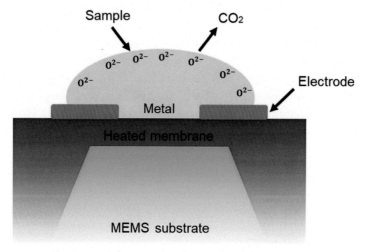

Figure 6.1 Schematic metal oxide-based CO_2 sensors.

6.6 Sol-gel CO_2 sensors

Sol-gel approach is one among the best in fabric technology, with its diversity of incorporating multiple materials at low value and simplified mechanism. It acts as the alternate technology for the conventional approach developing glass or other substances, with its operation at variable conditions. Sol-gel technique comprise of liquid to solid transition, encompass gelation, hydrolysis, and finally condensation (Dansby-Sparks et al., 2010; Lalam et al., 2020). The sol-gel based materials own its precise properties in delivering remarkable contribution toward gas sensing mechanisms. There are several benefits of making use of sol-gel method in gas sensing. Sol-gel method gives a highly sensitive and stable materials, which are produced by inducing the precursors and stabilizers. Moreover, extra strategies as doping and grafting add values to the material produced, which are desirable for a precise sensing system (Bahar et al., 2014; Nivens et al., 2002). Sol-gel based materials uphold a controlled nanoscale particle size and able to retain a thin layer on a sensing element. Sol-gel based thin films have shown remarkable advantages in sensing CO_2 gas (Mujahid et al., 2010).Sol-gel based sensors are commonly applied as pH indicators in identifying the CO_2 gas in hospital and environmental settings (Liu et al., 2017).

6.7 Polymer based CO_2 sensor

Polymer-based CO_2 sensors are mostly miniaturized version of the Severinghaus CO_2 probe. The advantage of polymer-based gas sensor over metal oxide sensor is that polymer films are capable of broadly detecting and identifying various constituents in the air in addition to target analyte. Hence, polymer sensors are multifunctional, and in general, polymer-based sensors are used in an array. Unlike other materials, polymers have vast versatility in its operating conditions, which can be tuned for excellent sensitivity and performance. The flexible polymers are able to operate effortlessly in high pressure and temperature environment. Moreover, it is cheap to synthesis and fine-tune polymers to attain the desirable working conditions (Waghuley et al., 2008). Polymer based CO_2 sensors are usually light in weight, which composed of compacted chemical and electronic structure. It enables it to fit for the portable CO_2 sensors (Molina et al., 2020). Polymers are modeled to capture the CO_2 concentration and transform the chemical signal to another physical signals such as absorption, resistance, waves, and other audible variables. Polymers are well established on CO_2

sensing electrode with suitable surface chemistry and doping elements, which enhanced the sensitivity and performance in detecting CO_2 and the related respiratory illness (Kazanskiy et al., 2021).

6.8 NASICON CO_2 sensors

NASICON powder ($Na_{2.8}Zr_2Si_{1.8}P_{1.2}O_{1.2}$), a metal oxide, has been proposed to be used as a promising solid electrolyte in sensing of CO_2. High ionic conductivity, high chemical stability, and the less influence of humidity on NASICON makes it a preferred material for CO_2 sensing (Zhang et al., 2021). The advantages of semiconducting metal oxide gas sensors over competing technologies such as conducting polymers, electrochemical, and quartz crystal are robustness and long–term stability (Yanase et al., 2021). Several researchers, including National Aeronautics and Space Administration (NASA), United States, have proposed that NASICON is the best sensing material for CO_2 sensors. Most commercially available microfabricated sensors (e.g., Cole–Palmer; Figaro) are made of a substrate heated by wire and coated with NASICON as a semiconducting film. The sensors rely on the changes of conductivity induced by the adsorption of gases and subsequent surface reactions. These micromachined sensors operate at a relatively high temperature of $200-450°C$, which results in significant power consumption. The sensors cannot function without small heaters (usually platinum or gold) on the back side to keep the sensors at operating temperatures (Zheng et al., 2018). Because of the high operating temperatures, NASICON sensors are inappropriate in potentially flammable environments or explosive environment such as grain stores. The output of the NASICON gas sensors varies logarithmically with the gas concentration. This limits the accuracy of the sensor and the overall measurement range of the sensor (Lorenc et al., 2015).

6.9 Colorimetric CO_2 sensors

Colorimetric gas sensors are one of preliminary type of detection strategy for identifying gas molecule with low cost of implementation and precise reading with low limit of detection. It is a visualized sensor, as the change of color denotes the detection results (Howes et al., 2005; Schmitt et al., 2016). Colorimetric sensors are also named as fluorescence sensors as it indicates the color changes based on the change in absorption wavelength, regarding the physiochemical changes in the analyte involved.

Mostly, colorimetric sensors were interpreted with the manipulation of pH of analyte. Colorimetric sensors are suitable for mobile type of sensors as they available in small size, effective energy conservation, inexpensive and portable or wireless type of sensor (Cho et al., 2021; Ko et al., 2020; Lin et al., 2018). With the advancement of wireless networking, colorimetric sensors can be integrated with waveguide sensitive gas sensor layer for acquiring simple signal transmission system. However, it is limited in use due to the low sensitivity caused by low limit absorption path with variant gas layer thicknesses. Colorimetric sensors are introduced for gas detection, especially for detecting toxic and hazardous gases, as the pH variation is significant and able to provide prompt reading with colorimetric sensors. Apart this, colorimetric gas sensors are implemented in medical applications, and one of its proficient functions is to detect CO_2 gas for monitoring asthma. Colorimetric sensor able to detect CO_2 from exhaled breath as the colorimetric probe is set up with pH sensitive chemical indicator. The fluorescence intensity changes with the use of fluorescence dye as the indicator CO_2 marker. Yet, the limitation in the dye suitability lowers its usage in colorimetric CO_2 sensor (Brown et al., 2016; Chu & Hsieh, 2019). The pH sensitive colorimetric CO_2 sensor is preferred over the fluorescence reference dyes due to its advantages in reducing noise interference, remote sensing, and electrical isolation. When CO_2 dissolves into water as aqueous bicarbonate the indicator responds to the change in pH of the aqueous layer supported within the sensor matrices. With the simple physiochemical change, colorimetric CO_2 sensor helps to detect CO_2 gas at ambient conditions. Some limitations of colorimetric CO_2 sensor is the lack of precision in regeneration and inability in giving digital image readout (Chatterjee & Sen, 2015).

6.10 Wet conductometric CO_2 sensors

The principal of wet conductometric sensor is based on the conductivity of liquids, which results from the dissociation of dissolved compounds. The sundering of target compounds induces the movement and migration of conductive ions in an electrolyte system, which developed an electrical field. As an external voltage source is supplied, a chaotic movement of ions takes place, which redirects the ions movement toward oppositely charged electrodes. Then, the electrolyte system neutralized as the ions are neutralized and expelled as neutral molecules (Li et al., 2010; Meng et al., 2014). CO_2 gas detected through wet conductometric sensor when the CO_2 gas dissociates in electrolyte and gives HCO_3^- and H_3O^+ ions. These changes the conductivity of the electrolyte system and the indicates the presence of CO_2 based on qualitative evaluations. Wet conductometric sensors for CO_2 detection is composed of a chamber containing electrolyte and two electrodes. The CO_2 gas able to diffuse into the chamber through a CO_2 permeable membrane fixed at the outer layer of the system. The CO_2 gas diffusion takes place, the impedance and electrical measurement of the dynamic conductometric are recorded and further analyzed (Akram et al., 2021; Di Francia et al., 2009). Wet conductometric sensing system works with an inexpensive theory, stable and easy to build set-up. However, the limitation is obvious in the accuracy of electrical measurement due to the highly possible contamination in electrolyte chamber. The contamination is common due to the external and unwanted compounds found in the distilled water. The presence of filter or resins at the outer layer of chamber may reduce the contamination but complicates the release and outgassing of CO_2 into chamber. As per demonstrated in literature reports, several calibration and drift controlling technique has been established to obtain accurate CO_2 gas measurement through wet conductometric sensors (Jaffrezic-Renault & Dzyadevych, 2008; Sharma et al., 2021).

6.11 Stow-Severinghaus electrode — ISFET CO_2 sensors

Stow-Severinghaus electrode was invented by Richard Stow and John Severinghaus, and that's where the electrode was named. Apart from many other inventions, Stow-Severinghaus electrode had special recognition due to the uniqueness in detection transcutaneous pCO_2. The experiment set-up is simple and similar to the wet-conductometric CO_2 sensor (Monk et al., 2021; Yue et al., 2010). Fig. 6.2 shows the set-up of Stow-Severinghaus electrode for CO_2 detection. The transition in the electrolyte pH is monitored by pH meter, which is plugged into the system and covered with thin membrane. The CO_2 dissociated in the electrolyte system is detected by the Stow-Severinghaus electrode, with the help of ion-impermeable and CO_2 permeable component. Polytetrafluoroethylene (PTFE) polymer is discovered as the suitable material for the rubber membrane, as the silicon material due to the better adhesion. Moreover, the electrolyte was fixed to sodium chloride (NaCl) solution with different concentration, as it more suitable for Stow-Severinghaus electrode, which improves the response time and sensitivity of the system (Hemming et al.,

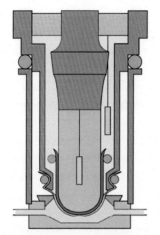
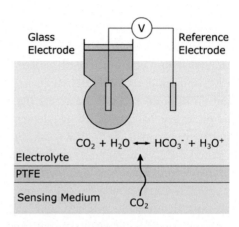

Figure 6.2 Schematic illustration of Stow-Severinghaus electrode set up. Left illustrates initial version of electrode discovered. Right illustrated the upgraded version with ISFET technology. *Reproduced with permission from Dervieux et al. (2021), Copyright 2021 MDPI.*

2016; Shitashima, 2010). Regarding system miniaturization, Stow-Severinghaus electrode CO_2 sensing is then upgraded to ion sensitive field effect transistor (ISFET). It is a type of metal oxide semiconductor, where the gate channel of the transistor is H_3O^+ sensitive. The excellence of this system is that it excludes the usage of pH meter. This is because the reference electrode added into the system mimics the pH meter function and indicate CO_2 detection. The rubber membrane is also upgraded to gas permeable membrane, as its suites bicarbonate ion dissociation in the electrolyte. ISFET sensor is preferred for small or laboratory scale CO_2 sensing due to the high accuracy in reading and stabilization of the set-up for good results (Umeda et al., 2021; Xie & Bakker, 2013).

6.12 Infrared CO$_2$ sensors

Various types of CO_2 sensors have evolved throughout the years, but Infrared (IR) radiation was the first to provide a substantial sensing method. Before 200 years, the science of infrared was introduced. In 1798, William Herschel, who discovered the IR spectrum, classified IR radiation as a light radiation beyond red light (Simon et al., 2022; Suleiman et al., 2022). IR radiation was first used in temperature measures based on radiometric measurements, with the introduction of various types of thermometers and bolometers (Karim & Andersson, 2013). The widespread adoption of smart technologies, as well as the rising demand for tele-technology, led in the development of infrared detectors in the late 20th century. The improvement in silicon technology has opened the door for the creation of infrared sensors that are combined with smart telecommunication technologies. With competing technologies such as photon sensors, Schottky diodes, quantum dots, and so on, the actual IR sensor breakthrough was noticed. This discovery has given IR gas detectors positive feedback in terms of manufacturing small, light, low-cost, and easy-to-handle IR sensors that are suited for a wide range of applications. IR-based gas sensors were first used to detect dangerous and combustible gaseous substances (Alghoul et al., 2022; Chen et al., 2022; Sakudo, 2016). The mechanism has now been expanded to include HVAC systems (heating, ventilation, and air conditioning), as well as air quality monitoring at various facilities. IR detectors are used in medical applications such as incubation, neurosurgery, breath analyzer, biomarker detection, and so on, because the severity of human pulmonary diseases is targeted based on CO_2 concentration (Deng et al., 2018; Schober & Schwarte, 2020). The peculiarity of IR gas sensors in achieving their specific absorption qualities distinguishes their superior performance for assessing CO_2 gas with good sensitivity, selectivity, repeatability, resolution, and hysteresis (Corsi, 2012; Popa & Udrea, 2019). There are numerous types of IR CO_2 sensors, depending on the material used and the technique used to convert electromagnetic radiation to electrical impulses. The nondispersive IR-based CO_2 gas sensor is the most appropriate and preferable sensing mechanism for detecting and monitoring CO_2 gaseous emissions (Vafaei et al., 2020).

6.12.1 NDIR-based CO_2 sensors

NDIR is basically a type of nondispersive infrared, where a beam of infrared (IR) light is emitted from a light source, which does not "disperse" or become scattered by substances between the light source and a detector (Hodgkinson et al., 2013). The most commercially available CO_2 sensors are infrared detectors. Infrared detectors produce quick response times and reliable. The advantages of infrared gas sensors over other analytical instruments are the cheaper cost, compact size, easy process control, easy mass production, and continuous measurement (Jha, 2021). NDIR CO_2 sensors are simpler in structure and easy to use. In the late 1970s, when IR and silicon semiconductor technology were at their peak, NDIR sensors gained popularity. NDIR-based gas sensors have acquired human trust in the medical and pharmaceutical industries in the last decade, as the technology has shown widespread interest and application in the environmental and manufacturing industries (Mao et al., 2021; Rolle & Sega, 2018). The introduction of high-performance semiconductors has transformed mechanical NDIR sensors into modern electronics NDIR sensors. Although the ideal performance and strategies of NDIR gas sensors have been reported, research to improve the mechanism and eliminate existing shortcomings cannot be denied. Due to its bulk structure, NDIR gas sensors are often subject to objections due to the installation of the system. In addition, the performance of the NDIR gas sensor has been questioned due to the presence of interferences in the gas detection system (Bley et al., 2015; Ch'ien et al., 2020; de Hoyos-Vazquez et al., 2019). The overlap of two different gas matrices or water vapor causes interference and reduces the accuracy of gas detection. The large absorption of water vapor occurs at $2-8$ μm, so the correction factor for removing/filtering water vapor is made prior to the NDIR gas measurement. Various bandpass and optical filters can be used to reduce such interference, but the ideal solution has not yet been determined. Multi-gas interference is common in NDIR gas chambers for detecting two different gases. The corrective action taken to overcome the interference is the use of interchannel interference constants coupled to the multi-optical filter. The IR detector is coupled to a bandpass filter from the interfering gas and the target channel is analyzed (Dinh et al., 2016). The working principle is based on the energy absorption characteristics of CO_2 in the infrared region. All gases selectively absorb infrared light energy, which corresponds to their own quantized vibrating energy. CO_2 absorbs infrared radiation at wavelengths of 2.7, 4.3, and 15 μm (Jia et al., 2019). The abundance of CO_2 in the enclosure is proportional to the amount of IR light absorbed. When the concentration of CO_2 increases, the more infrared light is absorbed, and less light is detected. After the detection of the CO_2 the concentration of CO_2 is calculated in parts per million (ppm) or as a percent (%) by a known mathematical equation, which is present in the enclosure (Diharja et al., 2019). A schematic of a simple NDIR gas sensor design to detect the CO_2 gas is given as Fig. 6.3. The infrared light source is installed at one end of an enclosed cell and the light detector is installed at the opposite end. The light detector has a filter, so it singularly detects light in the electromagnetic spectrum, which is inlaid with CO_2. There is an inlet and outlet lying between the detector and the light source to allow the CO_2 to move freely inside the NDIR cell.

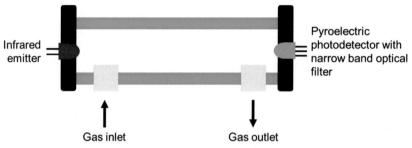

Figure 6.3 Schematic of NDIR gas chamber.

In NDIR based technology, the concentration of CO_2 gas is most measured by its absorption of infrared radiation within specific spectral range (Ch'ien et al., 2019; Cui et al., 2021). Usually most small gaseous molecules exhibit a vibrational mode that also lies in the mid-infrared (MIR) range of 2.5–25 µm, and related with stretching, twisting or bending their bonds. CO_2 gas has three vibrational modes: A symmetric stretch mode ($V_1 = 1350$ cm^{-1}), A bending mode ($V_2 = 670$ cm^{-1}) and an antisymmetric stretch mode ($V_3 = 2350$ cm^{-1}). The antisymmetric stretch mode corresponds with the wavelength 4.26 µm in MIR and this is the most useful wavelength for measuring CO_2 because there are only few molecules, which have a very little amount of significance of absorption at 4.26 µm range (Beduk et al., 2021; Popa & Udrea, 2019). The absorbance of CO_2 in MID IR range is presented in Fig. 6.4.

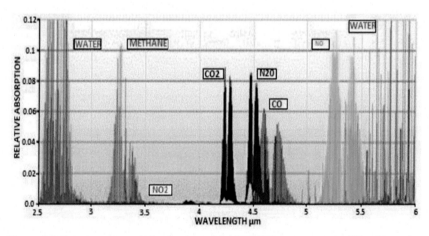

Figure 6.4 Relevant spectral distribution of the MID-IR light source, the CO_2 absorption bands and water vapor (Gibson & MacGregor, 2013).

Table 6.1 Comparison of commonly used CO_2 sensors.

Sensor	Operating voltage	Current (mA)	Range	Response time and accuracy	Warm up time (s)	Limitations
Sprint IR	3.2–5.0	15	0%–100% volume	0.05 s, ±70ppm ±5%	<60	Poorly related to cardiorespiratory condition, can't be used for long time monitoring, not internally calibrated (Scholz et al., 2015)
MG-811	6	200	0–10,000 ppm	<60 s	10,800	Not reliable and accurate in terms of result (Nurifhan et al., 2016)
MISIR	3.2–4.2	6	0–5000 ppm	120 s	<20	High response time and applicable for environmental use (Low Cost HVAC Carbon Dioxide Sensor NDIR CO2 Sensor – MisIR, n.d.)
MH410D	3.5–5.5	75–85	10%–100% volume	>30 s	90	Very high current, high warm up time and high response time (NDIR Infrared CO2 Gas Sensor – MH-410D, n.d.)

(Continued)

Table 6.1 Comparison of commonly used CO_2 sensors.—cont'd

Sensor	Operating voltage	Current (mA)	Range	Response time and accuracy	Warm up time (s)	Limitations
IR Prime 2	3—5	93	0%—10%	<30 s	<60	Very high current (Voltage Output Infrared Carbon Dioxide (CO2) Gas Sensor - Prime2, n.d.)
TGS4161	5	50	350—10,000 ppm	90 s	43,200	Very high current with high warm up time (Figaro USA Inc., 2005)
COZIR Ambient	3.2—5	<1.5	Ambient —5000 ppm	±50 ppm ±3% of reading	1.2	Applicable for environmental application only (SST CozIR–Blink – Low Power CO2 Sensor, n.d.)
Telaire 6613	5	19	Ambient —2000 ppm	±30 ppm ±5% of reading	<120	High current drawback, not appropriate for medical application (Telaire T6613, CO_2 Sensor Module, n.d.)

Table 6.2 Existing NDIR based CO_2 sensors for asthma monitoring.

Author	Method	Limitations
Yang et al. (2016)	Concentration and gas flow from same location, airway adapter, and dual channel detector.	Double channel sensor increases the complexity.
Yang et al. (2010)	NDIR, mainstream and sidestream both, concentration, IR Source (EMIRS200), Heimann Sensor GmbH, Dresden, Germany, temperature sensor.	Temperature increase. Temperature sensors produce heat and that may lead to an incorrect result.
Hodgkinson et al. (2013)	Low-cost injection molding technology, dual channel sensor, emitter (Gilway 1600), Pyroelectric detector from Perkin Elmer (PYS3228), NDIR, thermoplastic material with a gold reflective coating	Detector is costly Gold coating is expensive This design is still considered to be a short-term limit of detection for CO_2
Lee et al. (2016)	Thermopile detector, source ILT MR3–1089 (5V, 150 mA), 8629 CE and ADA4528-1 low noise amplifiers, ADuCM360 precision analog microcontroller, output of the sensor typically comes in an mV range, NDIR.	Could detect a large range of gases at the same time and it's not focused on CO_2 only Had the tendency to give wrong results in terms of CO_2 detection, which implies doubt using this device in medical sector.
Vincent and Gardner (2016)	MEMs based NDIR, portable breath analyser, thermopile detector, low power MEMS silicon on insulator (SOI) wideband IR emitter, lock-in amplifier, sidestream, and aluminum chamber.	Thermopile detectors are susceptible to high levels of noise and has been reported as the predominant noise source. Detection is slower than mainstream technique.
Rodr (2016)	A dual Pyroelectric sensor (LHi814 G2/G20 Perkin Elmer), modulation frequency of 250 mHz.	High response time typically 10 min. Needs AI to reduce response time This study is still under research.

(Continued)

Table 6.2 Existing NDIR based CO_2 sensors for asthma monitoring.—cont'd

Author	Method	Limitations
Zhao et al. (2014)	Colorimetric sensor, mainstream, response time 150 ms, Low detection limit of a few tens of ppm, which are suitable for IAQ monitoring, NDIR.	Delay due to aspiration by sampling tube (around 3 ft long).
Santoso and Setiaji (2013)	Capnography, side stream, successfully tested it during a laparoscopic surgery.	Another CO_2 sensor would be necessary to avoid false readings because the sensor was not internally calibrated with respect to changes in atmospheric pressure and temperature.
Scholz et al. (2015)	MID-IR LED based sensor, Acoustic based detector, MEMS microphone with a hermetically sealed small chamber containing CO_2 molecules, modulation frequency 1 kHz.	It produced huge noise and new strategies for signal generation were needed to reduce the influence of acoustic noise.

6.12.2 Incorporation of NDIR-CO_2 sensors for monitoring asthma

IR spectroscopy is the most recognized electromagnetic radiation technology used in the medical field. The defined IR wavelength and its light absorption contribute significantly to the diagnosis and treatment of diseases. Literature reviews show advances in IR spectroscopy in wound healing, rigidity, photodynamic therapy, neurological disorders, stem cell research, and psychometric disorders. Variety of IR based transducers have been developed and are being used widely to convert CO_2 concentration to electrical signal in asthma monitoring. NDIR based CO_2 sensors are the highly available sensors for detecting deviation on CO_2 gas for determining the asthmatic condition of a patient. Table 6.1 enlists several CO_2 sensors, along with their features, advantages, disadvantages, which are commercially available in the market. In various studies related to respiratory analysis, different ideas have been presented by different researchers. Studies have been performed with multiple strategies using IR– CO_2 sensors in medical research. The NDIR sensors are often incorporated with humidity, temperature, and air movement sensors to evaluate CO_2 gas by eliminating other disrupting factors and determine the triggered asthmatic condition of a subject. Table 6.2 shows the summary of studies conducted using NDIR sensor, which are being used in medical instrumentation industry. The ideal CO_2 monitoring for detecting asthma relies on the selection and performance of the CO_2 sensor. As discussed, NDIR based CO_2 sensors uphold its place top in the list for the most suitable CO_2 sensors, concerning the mass production and sensitivity in measuring the CO_2 expired gas at external settings. The advancement of NDIR based CO_2 sensors is yet to be implemented in determining asthmatic condition of human through capnography. Capnography, which measures expired CO_2 interprets the asthmatic condition into asthma, chronic obstructive pulmonary disease, pulmonary edema many more criteria, demands for a high performance NDIR and a sensitive system for rapid analysis and diagnosis. The use of NDIR based CO_2 sensor to design carbon dioxide measurement system for monitoring asthmatic condition is discussed in the next chapter.

References

Akram, R., Yaseen, M., Farooq, Z., Rauf, A., Almohaimeed, Z. M., Ikram, M., & Zafar, Q. (2021). Capacitive and conductometric type dual-mode relative humidity sensor based on 5,10,15,20-tetra phenyl porphyrinato nickel (II) (TPPNi). *Polymers, 13*(19). https://doi.org/10.3390/polym13193336

Alghoul, H., Farajat, F. Al, Alser, O., Snyr, A. R., Harmon, C. M., & Novotny, N. M. (2022). Intraoperative uses of near-infrared fluorescence spectroscopy in pediatric surgery: A. *Journal of Pediatric Surgery*. https://doi.org/10.1016/j.jpedsurg.2022.01.039

Aroutiounian, V. M. (2020). Metal oxide gas biomarkers of diseases for medical and health applications. *Biomedical Journal of Scientific & Technical Research, 29*(2), 22328−22336. https://doi.org/10.26717/bjstr.2020.29.004780

Bahar, M., Gholami, M., & Azim-Araghi, M. E. (2014). Sol-gel synthesized Titania nanoparticles deposited on porous polycrystalline silicon: Improved carbon dioxide sensor properties. *Materials Science in Semiconductor Processing, 26*(1), 491−500. https://doi.org/10.1016/j.mssp.2014.05.035

Beduk, T., Durmus, C., Hanoglu, S. B., Beduk, D., Salama, K. N., Goksel, T., Turhan, K., & Timur, S. (2021). Breath as the mirror of our body is the answer really blowing in the wind? Recent technologies in exhaled breath analysis systems as non-invasive sensing platforms. *TrAC − Trends in Analytical Chemistry, 143*, 116329. https://doi.org/10.1016/j.trac.2021.116329

Berman, E. S. F., Fladeland, M., Liem, J., Kolyer, R., & Gupta, M. (2012). Greenhouse gas analyzer for measurements of carbon dioxide, methane, and water vapor aboard an unmanned aerial vehicle. *Sensors and Actuators, B: Chemical, 169*, 128−135. https://doi.org/10.1016/j.snb.2012.04.036

Bley, T., Steffensky, J., Mannebach, H., Helwig, A., & Müller, G. (2015). Degradation monitoring of aviation hydraulic fluids using non-dispersive infrared sensor systems. *Sensors and Actuators, B: Chemical, 224*, 539−546. https://doi.org/10.1016/j.snb.2015.10.049

Blombach, B., & Takors, R. (2015). CO_2 − Intrinsic product, essential substrate, and regulatory trigger of microbial and mammalian production processes. *Frontiers in Bioengineering and Biotechnology, 3*(August), 1−11. https://doi.org/10.3389/fbioe.2015.00108

Brown, S. L., Goulsbra, C. S., Evans, M. G., Heath, T., & Shuttleworth, E. (2020). Low cost CO_2 sensing: A simple microcontroller approach with calibration and field use. *HardwareX, 8*, e00136. https://doi.org/10.1016/j.ohx.2020.e00136

Brown, M. K., Lazarus, D. V., Gonzales, S. R., Rich, W. D., Wozniak, M. J., Poeltler, D. M., & Katheria, A. C. (2016). Resistance of colorimetric carbon dioxide detectors commonly utilized in neonates. *Respiratory Care, 61*(8), 1003−1007. https://doi.org/10.4187/respcare.04507

Chaix, E., Guillaume, C., & Guillard, V. (2014). Oxygen and carbon dioxide solubility and diffusivity in solid food matrices: A review of past and current knowledge. *Comprehensive Reviews in Food Science and Food Safety, 13*(3), 261−286. https://doi.org/10.1111/1541-4337.12058

Chang, Y. Z., Lin, J. N., Li, S. D., & Liu, H. (2021). Adsorption of greenhouse gases (methane and carbon dioxide) on the pure and Pd-adsorbed stanene nanosheets: A theoretical study. *Surfaces and Interfaces, 22*, 100878. https://doi.org/10.1016/j.surfin.2020.100878

Chatterjee, C., & Sen, A. (2015). Sensitive colorimetric sensors for visual detection of carbon dioxide and sulfur dioxide. *Journal of Materials Chemistry A, 3*(10), 5642−5647. https://doi.org/10.1039/c4ta06321j

Chen, Z., Zeng, J., He, M., Zhu, X., & Shi, Y. (2022). Portable ppb-level carbon dioxide sensor based on flexible hollow waveguide cell and mid-infrared spectroscopy. *Sensors and Actuators B: Chemical, 359*(January), 131553. https://doi.org/10.1016/j.snb.2022.131553

Chien, P. J., Suzuki, T., Ye, M., Toma, K., Arakawa, T., Iwasaki, Y., & Mitsubayashi, K. (2020). Ultra-sensitive isopropanol biochemical gas sensor (Bio-sniffer) for monitoring of human volatiles. *Sensors (Switzerland), 20*(23), 1−13. https://doi.org/10.3390/s20236827

Ch'ien, L. B., Wang, Y. J., Shi, A. C., & Li, F. (2019). Wavelet filtering algorithm for improved detection of a methane gas sensor based on non-dispersive infrared technology. *Infrared Physics and Technology, 99*(January), 284–291. https://doi.org/10.1016/j.infrared.2019.04.025

Ch'ien, L. B., Wang, Y. J., Shi, A. C., Wang, X., Bai, J., Wang, L., & Li, F. (2020). Noise suppression: Empirical modal decomposition in non-dispersive infrared gas detection systems. *Infrared Physics and Technology, 108*(May), 103335. https://doi.org/10.1016/j.infrared.2020.103335

Cho, S. H., Suh, J. M., Eom, T. H., Kim, T., & Jang, H. W. (2021). Colorimetric sensors for toxic and hazardous gas detection: A review. *Electronic Materials Letters, 17*(1). https://doi.org/10.1007/s13391-020-00254-9

Chu, C.-S., & Hsieh, M.-W. (2019). Optical fiber carbon dioxide sensor based on colorimetric change of α-naphtholphthalein and CIS/ZnS quantum dots incorporated with a polymer matrix. *Optical Materials Express, 9*(7), 2937. https://doi.org/10.1364/ome.9.002937

Corsi, C. (2012). Infrared: A key technology for security systems. *Advances in Optical Technologies, 2012*(2). https://doi.org/10.1155/2012/838752

Cui, P., Zhao, J., Liu, M., Qi, M., Wang, Q., Li, Z., Suo, T., & Li, G. (2021). Non-invasive detection of medicines and edible products by direct measurement through vials using near-infrared spectroscopy: A review. *Infrared Physics and Technology, 115*(September 2020), 103687. https://doi.org/10.1016/j.infrared.2021.103687

Cummins, E. P., Selfridge, A. C., Sporn, P. H., Sznajder, J. I., & Taylor, C. T. (2014). Carbon dioxide-sensing in organisms and its implications for human disease. *Cellular and Molecular Life Sciences, 71*(5), 831–845. https://doi.org/10.1007/s00018-013-1470-6

Cummins, E. P., Strowitzki, M. J., & Taylor, C. T. (2020). Mechanisms and consequences of oxygen and carbon dioxide sensing in mammals. *Physiological Reviews, 100*(1), 463–488. https://doi.org/10.1152/physrev.00003.2019

Dansby-Sparks, R. N., Jin, J., Mechery, S. J., Sampathkumaran, U., Owen, T. W., Yu, B. D., Goswami, K., Hong, K., Grant, J., & Xue, Z. L. (2010). Fluorescent-dye-doped sol-gel sensor for highly sensitive carbon dioxide gas detection below atmospheric concentrations. *Analytical Chemistry, 82*(2), 593–600. https://doi.org/10.1021/ac901890r

Decker, M., Oelßner, W., & Zosel, J. (2019). Electrochemical CO_2 sensors with liquid or pasty electrolyte. *Carbon Dioxide Sensing*, 87–116. https://doi.org/10.1002/9783527688302.ch4

Deng, Y., Wang, Y., Zhong, G., & Yu, X. (2018). Simultaneous quantitative analysis of protein, carbohydrate and fat in nutritionally complete formulas of medical foods by near-infrared spectroscopy. *Infrared Physics and Technology, 93*(July), 124–129. https://doi.org/10.1016/j.infrared.2018.07.027

Dervieux, E., Théron, M., & Uhring, W. (2021). Carbon dioxide sensing—Biomedical applications to human subjects. *Sensors, 22*(188), 38–60. https://doi.org/10.1017/cbo9781139174176.006

Di Francia, G., Alfano, B., & La Ferrara, V. (2009). Conductometric gas nanosensors. *Journal of Sensors, 2009*. https://doi.org/10.1155/2009/659275

Diharja, R., Rivai, M., Mujiono, T., & Pirngadi, H. (2019). Carbon monoxide sensor based on non-dispersive infrared principle. *Journal of Physics: Conference Series, 1201*(1). https://doi.org/10.1088/1742-6596/1201/1/012012

Dinh, T. V., Choi, I. Y., Son, Y. S., & Kim, J. C. (2016). A review on non-dispersive infrared gas sensors: Improvement of sensor detection limit and interference correction. *Sensors and Actuators, B: Chemical, 231*, 529–538. https://doi.org/10.1016/j.snb.2016.03.040

Fajkus, M., Nedoma, J., Martinek, R., Vasinek, V., Nazeran, H., & Siska, P. (2017). A non-invasive multichannel hybrid fiber-optic sensor system for vital sign monitoring. *Sensors (Switzerland), 17*(1), 1—17. https://doi.org/10.3390/s17010111

Figaro USA Inc. (2005). *Tgs 4161 - for the detection of carbon dioxide* (pp. 1—2). http://www.figarosensor.com/products/4161pdf.pdf.

Fritzsche, E., Gruber, P., Schutting, S., Fischer, J. P., Strobl, M., Müller, J. D., Borisov, S. M., & Klimant, I. (2017). Highly sensitive poisoning-resistant optical carbon dioxide sensors for environmental monitoring. *Analytical Methods, 9*(1), 55—65. https://doi.org/10.1039/c6ay02949c

Ghorbani, R., & Schmidt, F. M. (2017). Real-time breath gas analysis of CO and CO_2 using an EC-QCL. *Applied Physics B: Lasers and Optics, 123*(5), 1—11. https://doi.org/10.1007/s00340-017-6715-x

Gibson, D., & MacGregor, C. (2013). A novel solid state non-dispersive infrared CO_2 gas sensor compatible with wireless and portable deployment. *Sensors (Switzerland), 13*(6), 7079—7103. https://doi.org/10.3390/s130607079

Hanafi, R., Mayasari, R. D., Masmui, Agustanhakri, Raharjo, J., & Nuryadi, R. (2019). Electrochemical sensor for environmental monitoring system: A review. *AIP Conference Proceedings, 2169*(November). https://doi.org/10.1063/1.5132657

Hemming, M., Kaiser, J., Heywood, K., Bakker, D., Boutin, J., Shitashima, K., Lee, G., Legge, O., & Onken, R. (2016). Measuring pH variability using an experimental sensor on an underwater glider. *Ocean Science Discussions*, 1—21. https://doi.org/10.5194/os-2016-78

Hodgkinson, J., Smith, R., Ho, W. O., Saffell, J. R., & Tatam, R. P. (2013). Non-dispersive infra-red (NDIR) measurement of carbon dioxide at 4.2 μm in a compact and optically efficient sensor. *Sensors and Actuators, B: Chemical, 186*, 580—588. https://doi.org/10.1016/j.snb.2013.06.006

Honeycutt, W. T., Kim, T., Ley, M. T., & Materer, N. F. (2021). Sensor array for wireless remote monitoring of carbon dioxide and methane near carbon sequestration and oil recovery sites. *RSC Advances, 11*(12), 6972—6984. https://doi.org/10.1039/d0ra08593f

Howes, D. W., Shelley, E. S., & Pickett, W. (2005). Colorimetric carbon dioxide detector to determine accidental tracheal feeding tube placement. *Canadian Journal of Anesthesia, 52*(4), 428—432. https://doi.org/10.1007/BF03016289

de Hoyos-Vazquez, F. F., Carreño-de León, M. C., Serrano-Nuñez, E. O., Flores-Alamo, N., & Solache Ríos, M. J. (2019). Development of a novel non-dispersive infrared multi sensor for measurement of gases in sediments. *Sensors and Actuators, B: Chemical, 288*(March), 486—492. https://doi.org/10.1016/j.snb.2019.03.017

Jaffrezic-Renault, N., & Dzyadevych, S. V. (2008). Conductometric microbiosensors for environmental monitoring. *Sensors, 8*(4), 2569—2588. https://doi.org/10.3390/s8042569

Jha, R. K. (2021). Non-dispersive infrared gas sensing technology: A review. *IEEE Sensors Journal, 22*(1), 6—15. https://doi.org/10.1109/jsen.2021.3130034

Jia, X., Roels, J., Baets, R., & Roelkens, G. (2019). On-chip non-dispersive infrared CO_2 sensor based on an integrating cylinder. *Sensors (Switzerland), 19*(19), 1—14. https://doi.org/10.3390/s19194260

Karim, A., & Andersson, J. Y. (2013). Infrared detectors: Advances, challenges and new technologies. *IOP Conference Series: Materials Science and Engineering, 51*(1). https://doi.org/10.1088/1757-899X/51/1/012001

Kazanskiy, N. L., Butt, M. A., & Khonina, S. N. (2021). Carbon dioxide gas sensor based on polyhexamethylene biguanide polymer deposited on silicon nano-cylinders metasurface. *Sensors (Switzerland), 21*(2), 1—14. https://doi.org/10.3390/s21020378

Khan, M. A. H., Rao, M. V., & Li, Q. (2019). Recent advances in electrochemical sensors for detecting toxic gases: NO_2, SO_2 and H_2S. *Sensors (Switzerland), 19*(4). https://doi.org/10.3390/s19040905

Ko, K., Lee, J. yeon, & Chung, H. (2020). Highly efficient colorimetric CO_2 sensors for monitoring CO_2 leakage from carbon capture and storage sites. *Science of the Total Environment, 729*, 138786. https://doi.org/10.1016/j.scitotenv.2020.138786

Kroukamp, O., & Wolfaardt, G. M. (2009). CO_2 production as an indicator of biofilm metabolism. *Applied and Environmental Microbiology, 75*(13), 4391–4397. https://doi.org/10.1128/AEM.01567-08

Lalam, N. R., Lu, P., Lu, F., Hong, T., Badar, M., & Buric, M. P. (2020). *Distributed carbon dioxide sensor based on sol-gel silica-coated fiber and optical frequency domain reflectometry (OFDR)*. https://doi.org/10.1117/12.2568653

Lee, B. R., Kester, W., & Seebeck, T. J. (2016). Complete gas sensor circuit using nondispersive infrared (NDIR). *Analog Dialogue, 50–10*(October), 1–9.

Li, D., Hu, J., Wu, R., & Lu, J. G. (2010). Conductometric chemical sensor based on individual CuO nanowires. *Nanotechnology, 21*(11).

Lin, C., Xian, X., Qin, X., Wang, D., Tsow, F., Forzani, E., & Tao, N. (2018). High performance colorimetric carbon monoxide sensor for continuous personal exposure monitoring. *ACS Sensors, 3*(2), 327–333. https://doi.org/10.1021/acssensors.7b00722

Liu, Y., Lee, D., Zhang, X., & Yoon, J. (2017). Fluoride ion activated CO_2 sensing using sol-gel system. *Dyes and Pigments, 139*, 658–663. https://doi.org/10.1016/j.dyepig.2016.12.069

Liu, L. L., Morgan, S. P., Correia, R., & Korposh, S. (2022). A single-film fiber optical sensor for simultaneous measurement of carbon dioxide and relative humidity. *Optics and Laser Technology, 147*(April 2021), 107696. https://doi.org/10.1016/j.optlastec.2021.107696

Lorenc, P., Strzelczyk, A., Chachulski, B., & Jasinski, G. (2015). Properties of Nasicon-based CO_2 sensors with $Bi_8Nb_2O_{17}$ reference electrode. *Solid State Ionics, 271*, 48–55. https://doi.org/10.1016/j.ssi.2014.12.002

Low cost HVAC carbon dioxide sensor NDIR CO_2 sensor — MisIR, (n.d.). https://www.isweek.com/product/low-cost-hvac-carbon-dioxide-sensor-ndir-co2-sensor-misir_18.html.

Mao, K., Xu, J., Jin, R., Wang, Y., & Fang, K. (2021). A fast calibration algorithm for nondispersive infrared single channel carbon dioxide sensor based on deep learning. *Computer Communications, 179*(August), 175–182. https://doi.org/10.1016/j.comcom.2021.08.003

Mendes, J. P., Coelho, L., Kovacs, B., de Almeida, J. M. M. M., Pereira, C. M., Jorge, P. A. S., & Borges, M. T. (2019). Dissolved carbon dioxide sensing platform for freshwater and saline water applications: Characterization and validation in aquaculture environments. *Sensors (Switzerland), 19*(24). https://doi.org/10.3390/s19245513

Meng, X., Kim, S., Puligundla, P., & Ko, S. (2014). Carbon dioxide and oxygen gas sensors-possible application for monitoring quality, freshness, and safety of agricultural and food products with emphasis on importance of analytical signals and their transformation. *Journal of the Korean Society for Applied Biological Chemistry, 57*(6), 723–733. https://doi.org/10.1007/s13765-014-4180-3

Mills, A. (2009). Optical sensors for carbon dioxide and their applications. *Sensors for Environment, Health and Security, 13*, 347–370.

Molina, A., Escobar-Barrios, V., & Oliva, J. (2020). A review on hybrid and flexible CO_2 gas sensors. *Synthetic Metals, 270*(October), 116602. https://doi.org/10.1016/j.synthmet.2020.116602

Monk, S. A., Schaap, A., Hanz, R., Borisov, S. M., Loucaides, S., Arundell, M., Papadimitriou, S., Walk, J., Tong, D., Wyatt, J., & Mowlem, M. (2021). Detecting and mapping a CO_2 plume with novel autonomous pH sensors on an underwater vehicle. *International Journal of Greenhouse Gas Control, 112*(September), 103477. https://doi.org/10.1016/j.ijggc.2021.103477

Mujahid, A., Lieberzeit, P. A., & Dickert, F. L. (2010). Chemical sensors based on molecularly imprinted sol-gel materials. *Materials, 3*(4), 2196—2217. https://doi.org/10.3390/ma3042196

NDIR. *Infrared CO_2 Gas Sensor — MH-410D*, (n.d.). https://www.isweek.com/product/ndir-infrared-co2-gas-sensor-mh-410d_143.html.

Neethirajan, S., Freund, M. S., Jayas, D. S., Shafai, C., Thomson, D. J., & White, N. D. G. (2010). Development of carbon dioxide (CO_2) sensor for grain quality monitoring. *Biosystems Engineering, 106*(4), 395—404. https://doi.org/10.1016/j.biosystemseng.2010.05.002

Neethirajan, S., Jayas, D. S., & Sadistap, S. (2009). Carbon dioxide (CO_2) sensors for the agrifood industry — A review. *Food and Bioprocess Technology, 2*(2), 115—121. https://doi.org/10.1007/s11947-008-0154-y

Nivens, D. A., Schiza, M. V., & Angel, S. M. (2002). Multilayer sol-gel membranes for optical sensing applications: Single layer pH and dual layer CO_2 and NH_3 sensors. *Talanta, 58*(3), 543—550. https://doi.org/10.1016/S0039-9140(02)00323-5

Nurifhan, A., B, M., Prakash, O., & Ahmed, S. (2016). Portable respiratory CO_2 monitoring device for early screening of asthma. *International Conference on Advances in Computing, Electronics and Communication, December*, 90—94. https://doi.org/10.15224/978-1-63248-113-9-61

Ogawa, K., Koyama, S., Ishizawa, H., Fujiwara, S., & Fujimoto, K. (2018). Simultaneous measurement of heart sound, pulse wave and respiration with single fiber Bragg Grating sensor. In *MeMeA 2018 — 2018 IEEE International Symposium on medical measurements and applications, Proceedings*. https://doi.org/10.1109/MeMeA.2018.8438629

Oprea, A., Degler, D., Barsan, N., Hemeryck, A., & Rebholz, J. (2018). Basics of semiconducting metal oxide-based gas sensors. In *Gas sensors based on conducting metal oxides: Basic understanding, technology and applications*. Elsevier Inc. https://doi.org/10.1016/B978-0-12-811224-3.00003-2

Popa, D., & Udrea, F. (2019). Towards integrated mid-infrared gas sensors. *Sensors (Switzerland), 19*(9), 1—15. https://doi.org/10.3390/s19092076

Puligundla, P., Jung, J., & Ko, S. (2012). Carbon dioxide sensors for intelligent food packaging applications. *Food Control, 25*(1), 328—333. https://doi.org/10.1016/j.foodcont.2011.10.043

Quan, V. M., Sen Gupta, G., & Mukhopadhyay, S. (2011). Review of sensors for greenhouse climate monitoring. In *SAS 2011 — IEEE sensors applications symposium, proceedings* (pp. 112—118). https://doi.org/10.1109/SAS.2011.5739816

Rodr, G. A. (2016). CO_2 measurement system based on pyroelectric detector. *SciELO Analytics, 62*(June), 278—281.

Rolle, F., & Sega, M. (2018). Carbon dioxide determination in atmosphere by non dispersive infrared spectroscopy: A possible approach towards the comparability with seawater CO_2 measurement results. *Measurement: Journal of the International Measurement Confederation, 128*(July), 479—484. https://doi.org/10.1016/j.measurement.2018.07.007

Sakudo, A. (2016). Near-infrared spectroscopy for medical applications: Current status and future perspectives. *Clinica Chimica Acta, 455*, 181—188. https://doi.org/10.1016/j.cca.2016.02.009

Santoso, D., & Setiaji, F. D. (2013). Design and implementation of capnograph for laparoscopic surgery. *International Journal of Information and Electronics Engineering, 3*(5), 523—528. https://doi.org/10.7763/ijiee.2013.v3.370

Schmitt, K., Tarantik, K., Pannek, C., Sulz, G., & Wöllenstein, J. (2016). Colorimetric gas sensing with enhanced sensitivity. *Procedia Engineering, 168*, 1237—1240. https://doi.org/10.1016/j.proeng.2016.11.430

Schober, P., & Schwarte, L. A. (2020). Thinking out of the (big) box: A wearable near-infrared spectroscopy monitor for the helicopter emergency medical service. *Air Medical Journal, 39*(2), 120−123. https://doi.org/10.1016/j.amj.2019.10.002

Scholz, L., Perez, A. O., Knobelspies, S., Wöllenstein, J., & Palzer, S. (2015). MID-IR led-based, photoacoustic CO_2 sensor. *Procedia Engineering, 120*, 1233−1236. https://doi.org/10.1016/j.proeng.2015.08.837

Sharma, B., Karuppasamy, K., Vikraman, D., Jo, E. B., Sivakumar, P., & Kim, H. S. (2021). Porous, 3D-hierarchical α-$NiMoO_4$ rectangular nanosheets for selective conductometric ethanol gas sensors. *Sensors and Actuators B: Chemical, 347*(March), 130615. https://doi.org/10.1016/j.snb.2021.130615

Shitashima, K. (2010). Evolution of compact electrochemical in-situ pH-pCO_2 sensor using ISFET-pH electrode. *MTS/IEEE Seattle, OCEANS, 2010*(October), 2−6. https://doi.org/10.1109/OCEANS.2010.5663782

Siefker, Z. A., Hodul, J. N., Zhao, X., Bajaj, N., Brayton, K. M., Flores-Hansen, C., Zhao, W., Chiu, G. T. C., Braun, J. E., Rhoads, J. F., & Boudouris, B. W. (2021). Manipulating polymer composition to create low-cost, high-fidelity sensors for indoor CO_2 monitoring. *Scientific Reports, 11*(1), 1−10. https://doi.org/10.1038/s41598-021-92181-4

Simon, J., Tsetsgee, O., Iqbal, N. A., Sapkota, J., Ristolainen, M., Rosenau, T., & Potthast, A. (2022). A fast method to measure the degree of oxidation of dialdehyde celluloses using multivariate calibration and infrared spectroscopy. *Carbohydrate Polymers, 278*(November 2021), 118887. https://doi.org/10.1016/j.carbpol.2021.118887

SST CozIR-Blink-low power CO_2 sensor. ((n.d.).

Suleiman, M., Abu-aqil, G., Sharaha, U., & Riesenberg, K. (2022). Enables early determination of *Pseudomonas aeruginosa*'s susceptibility to. *Spectrochimica Acta A: Molecular and Biomolecular Spectroscopy.* https://doi.org/10.1016/j.saa.2022.121080, 121080.

Tang, K. H., Tang, Y. J., & Blankenship, R. E. (2011). Carbon metabolic pathways in phototrophic bacteria and their broader evolutionary implications. *Frontiers in Microbiology, 2*(August), 1−23. https://doi.org/10.3389/fmicb.2011.00165

Telaire T6613, CO_2 Sensor Module. (n.d.). https://www.amphenol-sensors.com/en/telaire/co2/525-co2-sensor-modules/321-t6613.

Umeda, A., Ishizaka, M., Ikeda, A., Miyagawa, K., Mochida, A., Takeda, H., Takeda, K., Fukushi, I., Okada, Y., & Gozal, D. (2021). Recent insights into the measurement of carbon dioxide concentrations for clinical practice in respiratory medicine. *Sensors, 21*(16). https://doi.org/10.3390/s21165636

Vafaei, M., Amini, A., & Siadatan, A. (2020). Breakthrough in CO_2 measurement with a chamberless NDIR optical gas sensor. *IEEE Transactions on Instrumentation and Measurement, 69*(5), 2258−2268. https://doi.org/10.1109/TIM.2019.2920702

Vajhadin, F., Mazloum-Ardakani, M., & Amini, A. (2021). Metal oxide-based gas sensors for the detection of exhaled breath markers. *Medical Devices & Sensors, 4*(1), 1−11. https://doi.org/10.1002/mds3.10161

Vijayakumari, A. M., Oraon, A. R., Ahirwar, S., Kannath, A., Suja, K. J., & Basu, P. K. (2021). Defect state reinforced microwave-grown $CuxO/NiO$ nanostructured matrix engineered for the development of selective CO_2 sensor with integrated micro-heater. *Sensors and Actuators B: Chemical, 345*(July), 130391. https://doi.org/10.1016/j.snb.2021.130391

Vincent, T. A., & Gardner, J. W. (2016). A low cost MEMS based NDIR system for the monitoring of carbon dioxide in breath analysis at ppm levels. *Sensors and Actuators, B: Chemical, 236*, 954−964. https://doi.org/10.1016/j.snb.2016.04.016

Voltage Output Infrared Carbon Dioxide (CO_2) Gas Sensor − Prime2, (n.d.). https://www.isweek.com/product/voltage-output-infrared-carbon-dioxide-co2-gas-sensor-prime2_26.html.

Waghuley, S. A., Yenorkar, S. M., Yawale, S. S., & Yawale, S. P. (2008). Application of chemically synthesized conducting polymer-polypyrrole as a carbon dioxide gas sensor. *Sensors and Actuators, B: Chemical, 128*(2), 366−373. https://doi.org/10.1016/j.snb.2007.06.023

Xie, X., & Bakker, E. (2013). Non-severinghaus potentiometric dissolved CO_2 sensor with improved characteristics. *Analytical Chemistry, 85*(3), 1332−1336. https://doi.org/10.1021/ac303534v

Yanase, I., Hayashizaki, K., Kakiage, M., & Takeda, H. (2021). Novel application of Tb-substituted layered double hydroxides to capturing and photoluminescence detecting CO_2 gas at ambient temperature. *Inorganic Chemistry Communications, 125*(October 2020), 108394. https://doi.org/10.1016/j.inoche.2020.108394

Yang, J., An, K., Wang, B., & Wang, L. (2010). New mainstream double-end carbon dioxide capnograph for human respiration. *Journal of Biomedical Optics, 15*(6), 065007. https://doi.org/10.1117/1.3523620

Yang, J., Chen, B., Burk, K., Wang, H., & Zhou, J. (2016). A mainstream monitoring system for respiratory CO_2 concentration and gasflow. *Journal of Clinical Monitoring and Computing, 30*(4), 467−473. https://doi.org/10.1007/s10877-015-9739-y

Yoo, W. J., Jang, K. W., Seo, J. K., Heo, J. Y., Moon, J. S., Park, J. Y., & Lee, B. (2010). Development of respiration sensors using plastic optical fiber for respiratory monitoring inside MRI system. *Journal of the Optical Society of Korea, 14*(3), 235−239. https://doi.org/10.3807/JOSK.2010.14.3.235

Yue, Z., Niu, W., Zhang, W., Liu, G., & Parak, W. J. (2010). Detection of CO_2 in solution with a Pt−NiO solid-state sensor. *Journal of Colloid and Interface Science, 348*(1), 227−231. https://doi.org/10.1016/j.jcis.2010.04.020

Zhang, B., Ma, K., Lv, X., Shi, K., Wang, Y., Nian, Z., Li, Y., Wang, L., Dai, L., & He, Z. (2021). Recent advances of NASICON-$Na_3V_2(PO_4)_3$ as cathode for sodium-ion batteries: Synthesis, modifications, and perspectives. *Journal of Alloys and Compounds, 867*, 159060. https://doi.org/10.1016/j.jallcom.2021.159060

Zhao, D., Miller, D., Xian, X., Tsow, F., & Forzani, E. S. (2014). A novel real-time carbon dioxide analyzer for health and environmental applications. *Sensors and Actuators, B: Chemical, 195*, 171−176. https://doi.org/10.1016/j.snb.2013.12.110

Zheng, Q., Yi, H., Li, X., & Zhang, H. (2018). Progress and prospect for NASICON-type $Na_3V_2(PO_4)_3$ for electrochemical energy storage. *Journal of Energy Chemistry, 27*(6), 1597−1617. https://doi.org/10.1016/j.jechem.2018.05.001

CHAPTER SEVEN

Design of carbon dioxide measurement device for asthma monitoring

With reference to the details shared in previous chapters, this chapter is on the design of carbon dioxide (CO_2) measuring tool to monitor asthmatic conditions. The presented work emphasizes the design of capnography tool using high performance infrared CO_2 sensor and low-cost electronic components integrated with algorithms for extraction of capnogram features for monitoring asthmatic condition of a patient. With reference to the previous chapters, this chapter presents the design of capnography in detection and monitoring asthma. The design involves utilization of CO_2 sensor and the computer-based algorithm for the acquisition of CO_2 gas and classification of asthma and nonasthma. The capnograph device detects the exhaled CO_2; generates capnogram waveform; assess the capnogram features and disclose the asthmatic condition. Hereby, the developed strategy discriminates the presence and absence of asthma based on the capnogram waveform features. The capnography device generated is able to display capnogram waveform and the numerical values in response to the breathing condition of a subject, which have shown an appealing interest for rapid diagnosis of asthmatic conditions.

7.1 Respiratory gas monitoring standard: ISO 21647

A regulatory standard is a compulsory aspect in every clinical and healthcare facility. Merriam-Webster defined standard as the something established by authority, custom, or a general consent, which stands as the model for a defined state of circumstances (Castillo-Martinez et al., 2021; Shaw & Lin, 2021). Unlike other field of applications, clinical, medical, and healthcare facilities have greater concern in attaining and practicing the governed standards to ensure human life is protected from any hazards (de Andrade e Silva et al., 2022). Globally recognized clinical organizations developed the regulatory standards as the guidelines of clinical practices. American Association for Respiratory Care and American Society of

Systems and Signal Processing of Capnography as a Diagnostic Tool for Asthma Assessment
ISBN: 978-0-323-85747-5
https://doi.org/10.1016/B978-0-323-85747-5.00009-7

Anesthesiologist are the example organizations generating standards for handling medical facilities and patients (Johri & Kumar, 2021). Governmental agents and standard developing agents generate standards based on the circumstances encountered and reported to the agencies (Hosseini Teshnizi et al., 2021). Food and Drug Administration (FDA) is the most recognized government agent, which establish standards for global needs. International Organization for Standardization (ISO), British Standard Institutes (BSI), American Society for Testing and Materials (ASTM), International Electrotechnical Commission (IEC), and Committee European Normalization (CEN) are examples of nongovernmental standard developing agents (An et al., 2022; Nguyen et al., 2021). Among these, ISO is the mainly referred and established standard in multiples fields. Medical and healthcare field upholds many ISO standards, which are specifically established for a specific regulatory.

In this chapter, ISO 21647 standard for respiratory gas monitors (RGMs) is discussed. ISO 21647 is a standard developed for medical electrical equipment, particularly focuses on the requirements of basic safety and the essential performances of an RGM (Hosseini Teshnizi et al., 2021; Martini et al., 2018). RGM is clarified as a medical electrical equipment, which functions for measuring the gas levels such as gas volumes and partial pressures. RGM includes oxygen monitoring, carbon dioxide monitoring and anesthetic gas monitoring. RGM is used with an accessory, and a detecting agent, which acts as the sampling tool for gas measurements. The accessories of an RGM are designated by the manufacturer, as per the ISO 21,647 standard (Onyancha et al., 2021). The standard defines the regulatory for RGM and the accompanying accessories, as it directly handled by medical personnel and patients. ISO 21647 is closely associated with the International Electrotechnical Commission (IEC) standards (Ibrahim et al., 2022). IEC standards are addressed with RGM due to the installed electrical components and sensors for the continuous gas measuring. An RGM is allowed to be used by humans once the medical device is established with ISO 21674 and the related IEC standards as the basis for the equipment use and operations (Young & Schmid, 2018).

7.2 Current standard for respiratory gas monitoring

The previous standards for RGM are replaced with ISO/IEC standards, which are well-established for current RGM purposes (Deacon & Pratt, 2021). The combination of both standards represents the merge of standards and its regulatory from the individual RGM tools such as anesthetic monitoring tools, oxygen measuring device, and carbon dioxide measuring tools. The current RGM standard is established for manufacturers for ensuring the standard of all RGM applied in clinical facilities are the same. Moreover, customers and users of RGM have regulatory of fail-proof system to evaluate the accepting or rejecting of latest RGM devices (Keijzer & Scheeren, 2021; Pella et al., 2018). For users, mainly the clinicians are advised to understand the standard in detail for affirming an RGM device application, based on safety and performances, in comparison to the other RGM devices. In this chapter, the established ISO21647 standard for capnography is discussed in the aim of developing high performance real-time carbon dioxide monitoring device.

7.2.1 Capnography calibration

The frequency of calibration of a capnography tool determines the stability and the lifetime of the device. RGM with quantitative measurement requires standard calibration module for ensuring accurate outcomes and reliable monitoring (de Andrade et al., 2022; Sloop et al., 2022). Automotive calibration system requires time settings to ensure no interruption occurs during a continuous monitoring. Automatic zeroing is highly restricted to be applied in capnography tool. It causes disruption to the inspiration phase, where the amount of CO_2 gas is present in a small volume. Auto-setting may affect the CO_2 sensor and produce inaccurate reading during the expiration phase. Moreover, automatic zeroing may cause breakdown in the CO_2 absorber and the valves during continuous monitoring. Hence, a standardized procedure for calibration has been established for RGM tools (Wu et al., 2022). A detailed and long-term calibration standard is prepared with long steps of procedure and with infrequency time duration. In addition, short and brief calibration procedures are established to calibrate the capnography tool prior to every usage. With the current optical technology, infrared sensors are often utilized for developing CO_2 sensors. Infrared sensors able to operate without calibration for a minimum of 30 days, perhaps longer than 1 month, which are determined by the CO_2 sensor set-up and performances (Li et al., 2022). The latest technology has provided sensors, which eliminates calibration for at least a year. Apart calibration, there are several other techniques to determine the accuracy of the capnography. The measurements of end-tidal CO_2 (etCO$_2$) may indicate the accuracy of a device values (Chee et al., 2021). A capnography instrument may result in invalid $etCO_2$ reading as the accuracy of the instrument deteriorates. The commonly applied equipment such as beam occluder, and optical filter ensures that the signal and outcomes provided by capnography tool is valid for clinical applications (Olarte et al., 2016).

7.2.2 Electromagnetic interferences and radiation

Electromagnetic interferences and radiation rays are major factors, which interrupt the measuring system of capnography. Noise is a common name for the unwanted interferences. Such interferences affect the accuracy of gas measurements as well as impacts on the gas analysis (Paniccia et al., 2014; Wilson, 2021). Error generated in the gas analysis due to noise may result in obscure of signal and pattern, which determines the outcome. The effect of noise in the measurement system is classified as internal electromagnetic interferences. External electromagnetic radiation interferences as in radio frequency range also interrupt the capnography performances. The noises are transmitted in a narrow band, from the origin sources such as radio transmitters, power lines and oscillators (He et al., 2012). Radiation sourced from stratosphere is considered as wide-band radiations. Charges carried by electrical equipment such as pump, lighting, machines, and motors may generate the noise in system (Wu & Li, 2010). Antennas are electrical conductor agent in a gas measuring device, which receives the signal and the external noise interferences. With further analysis, the method of overcoming interferences is declared and included in the RGM standards (Ravindran et al., 2008). A proper shielding of an electrical capnography device is mandatory to ensure the critical electrical conductors are enclosed in a conductive material, which is grounded. Hence, validation test has been established to test the electromagnetic interferences and radiation on their immunity toward both internal and external interferences. Moreover, the system must be assured from electromagnetic interferences transmission, which may affect other nearby electrical equipment (de Carellán Mateo & Casamián-Sorrosal, 2022). ISO 21647 and IEC 6061 are the collateral standard for the developed standard and test.

7.3 ISO 21647 based capnography performance measures

The safety and performance of medical equipment is given the highest attention on developing the equipment standards. The standards do not determine the on the operation procedure of medical equipment and the location where the equipment will be placed or used. The standard for operation procedures and the external domains are established in the clinical practice standards for each designated institution (Shaban et al., 2022). Equipment standards state the step of functions of medical equipment, regardless of the user and the clinical condition. This standard is mandatory to be adhered by manufacturers and authorized regulatory of medical equipment, especially capnography. The FDA in the United States has adopted a consensus approach as the premise for premarket certification of medical equipment (Santos et al., 2016). There are several criteria for measuring the performances of capnography, which are established in the safety and performance of medical equipment standard. The criteria are discussed in the below section.

7.3.1 Measurement accuracy

The accuracy of RGM medical equipment is validated through the ability of the equipment in determining the respiratory gas mixture compositions (He et al., 2021; Jabłoński et al., 2011). Although it looks straightforward, there are many complicating factors involved in attaining the 99% accuracy, which is relatively the most difficult task for the medical equipment manufacturers. The accuracy measurement for respiratory gaseous is tedious to comply due to the mixture of the gaseous, which is not binary, or tertiary. It is a mixture of major respiratory gaseous, and other minor gaseous such as ethanol, methane, acetone, and water vapor. Among these, water vapor is the most disruptive agent for attaining the maximum accuracy of gas monitoring device till the current technology (Djerdj et al., 2021). Sidestream capnography and other RGM tools are preferred than mainstream system due to the susceptibility of sidestream to the interferences caused by water vapor. Water vapor in gas sampling may condense and deposit in the sampling tube during measurement. The deposition may cause contamination and disruption to the gas flow and result in accurate gas measurements (Djerdj et al., 2021; Mimoz et al., 2012). Among many types of gas analyzers, O_2 analyzer is the most simplified and easy handling device. The technology of measuring only the inspired O_2 is abundant and available in cheap prices. O_2 analyzers operate with the breath-by-breath analysis. Regardless of the technology implemented, the standard has specified the measurement accuracy for respiratory monitoring gas, as shown in Table 7.1 (Narimani et al., 2022). The primary objective of attaining maximum measurement frequency is the elimination of interferences effect caused by other gases in respiratory gas

Table 7.1 Measurement accuracy of RGM based on ISO 21647.

Gas	Measurement accuracy as in gas levels in % volume fraction
O_2	± (volume fraction of 2.5% + 2.5% gas level)
CO_2	± (volume fraction of 0.43% + 8% gas level)
Nitrous oxide	± (volume fraction of 2% + 8% gas level)
Halogenated agent	± (volume fraction of 0.2% + 15% gas level)

sampling. In the aim of achieving the maximum measurement accuracy and meet the established standards, a comprehensive testing tools and procedures are developed. The testing equipment represents the clinical condition, which helps to validate the gas monitoring equipment prior to the real clinical settings (Lermuzeaux et al., 2016). Among all, CO_2 measurement accuracy is the most challenging. After several modifications, ISO 21647 and American Society for Testing and Materials (ASTM) F-1456-01 has established that CO_2 reading shall be 12% of the test gas or 4 mmHg whichever is greater, over the full measurement rage of the capnometer, with one atmospheric pressure (Jabłoński et al., 2011).

7.3.2 Response time

Total response time is a significant criterion to be considered by manufactures and clinicians. The standard has established an RGM system has achieved total response time when the time it takes for a phase shift variation in gas levels at a sampling site to reach 90% of a final gas measurement. The total response time is important for clinicians to identify the abnormality in patient condition upon measurement (Chin et al., 2018). The response time is classified into rise time and delay time. Rise time is generally the lengthy and exponential response, assuming from 0% to 100%. However, it is categorized in many ways. Generally, the time taken by the equipment to attain 100% reading is not focused by clinicians. Fig. 7.1 shows the total response time, which is equal to the rising time and delay time (De Battista et al., 2019; Jashnsaz et al., 2021). The ideal response characteristics are determined by influencing the system with an instantaneous change, followed by

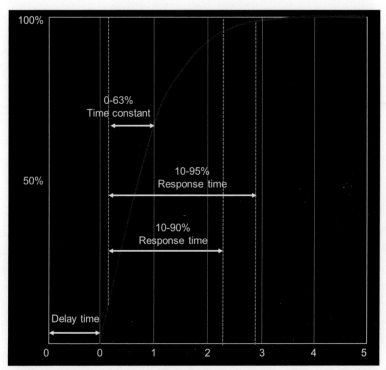

Figure 7.1 Illustration of total response time, which is equal to the delay and transit time of gas sampling through a sampling tube.

measuring the response time. Such procedure is established in the standard as step change. Delay time refers to the lagging time of the measurement system. Factors as length and diameter of sampling tube, sampling rate and the sample humidity and viscosity affects the delay time. Low viscosity samples with short and fast sampling tube and flow rate, respectively significantly reduces the delay time (Zhen & Xu, 2013). Manufacturers are obligated to provide clinicians with the entire response time so that they may make valid comparisons between different devices. Nondiverting capnograph has no delay time, which adds the value for the device to be used in medical facilities. Nondiverting capnography, which also known as mainstream capnography system determines the total response time based on the rapidity of gas flowing toward the sensor (Li et al., 2020; Ni et al., 2021). The standard governs gas flow at a particular velocity so that manufacturers with varying sample size cells can compete on total response time on an equal ground. Manufacturers are aware sidestream capnography requires a list of response time, which are specified for adults and children. Hence, clinicians can practice the standard usage by identifying the patient abnormalities based on the response time (Boehler et al., 2012). Response time is typically has given high attention in developing high accuracy capnography tools due to its effect in measuring the precise end-tidal and inspiratory gas volumes. Moreover, the response time is important for ensuring accurate capnogram waveform generation, which simultaneously indicates the gas concentration. Clinicians make interpretation based on the capnogram waveform monitored on the capnography tool, which emphasizes the importance of response time in capnography measurements (Zhevnenko et al., 2021).

7.3.3 Measurement drift

RGM devices are known for long period time of testing (Sarwar et al., 2021). The device will be monitored for more than few weeks to ensure the stability of device performance. Similarly, capnography devices are left for testing for an extended period. Capnography as an electronic tool consumes an interval of time to warm up after the power is turn on. The warmup period is mandatory for capnography prior to the use to ensure accurate measurement. The calibration set of the capnography may fall out of the range when the device is left for operation for a long period. Such conditions are defined as measurement drift (Maho et al., 2022). Capnography tool are designed to operate at ideal condition for a reasonable duration of time. Frequent calibration or recalibration for a short period of time (1 h) is not suitable with the available capnography technology. Hence, the regulatory authorities have considered measurement drift as a significant issue to be considered and the standard for measurement drift has been established (Agrahari & Singh, 2021; Lewis et al., 2015). The standard has emphasized that a capnography monitor is allowed to operate at least 6 h, considering the normal operations and settings. The device stability and accuracy are tested after 6 h, and calibration step may perform if there are any flaws in the capnography performance (Speers et al., 2022). With this, the measurement drift can be eliminated, and the stability of capnography tool is ensured.

7.3.4 Gas and vapor interference

The quantitative impacts of interfering gases on gas readings are usually mentioned in a gas measuring tool's operating manual. Table 7.2 shows the level of gas and vapor interference established in the standard prior to the RGM manufacture and operation. A capnography tool is usually

Table 7.2 Interfering gas and vapor level established in standard for testing.

Gas/vapor	Gas test level, ±20% volume fraction of specified level
Helium	50%
Xenon	80%
Isoflurane	5%
Enflurane	5%
Nitrous oxide	60%
Halothane	4%

vulnerable for unexpected gas interferences, mainly the therapeutic and anesthetic gases (Dinh et al., 2016). At these circumstances, manufacturers are instructed to disclose the extend of these gas interferences on the performance of a developed capnography tool for measuring the CO_2 gas. The extended gas and vapor interferences may cause inaccurate high and low readings. Patients with different illnesses may cause different types of gas interferences. Diabetic ketoacidosis may cause ethanol and acetone gas interference upon measuring CO_2 using capnography. Bowel obstruction may result in the release of methane gas in the respiratory gaseous, which interfere with capnography readings (Lin et al., 2022; Shevchenko et al., 2021). Moreover, isopropanol gas originated from the alcohol wipes in the medical applications and equipment also causes interference effect to the capnography measurement. During the cleaning of medical equipment, capnography tools may accidently sample vapors of the alcohol, resulting in the extended faultiness of capnography tool, depending on the technology implemented (Duncan & Pratt, 2021). Besides, the commonly used stabilizing gases in medical facility may cause interference in the capnography measurements. O_2 and helium, known as heliox is commonly used for patients suffering from upper respiratory obstruction (Mettelman et al., 2022). Anesthetic vapor such as Xenon are being in use for research purposes, which result in major gas interferences issue to the capnography measurement (Fleming et al., 2018; Regazzi et al., 2022). Hence, the established standard for RGM specified the extend of gas and vapor interference for ensuring the highly accurate CO_2 gas measurement prior to the medical applications.

7.3.5 Signal processing considerations

RGM tools prominently detect the gas and transform into signal, which are usually displays in a waveform pattern. Like other medical equipment such as electrocardiogram (ECG), capnography precision is mainly determined through the waveform pattern (Fuss et al., 2019). Moreover, the waveform is further analyzed to determine the accuracy of CO_2 measurements. Identification of possible errors and its sources are the main step of performing signal processing. Any uncertainty recognized in the capnography outcome is related to signal processing. The error in the signal processing may not occur due to a single source of problem (Huang & Wang, 2021). However, an error in signal processing may result from various sources of the capnography setting such as calibration, measurement, interface, sampling, and external domain conditions such as humidity, pressure, and temperature. The established standards have emphasized the importance of solving the uncertainty in signal processing, prior to the accurate capnography performance (Benedetto & Tosti, 2017; Charlton et al., 2021). The quality of CO_2 measurement is denoted based on the improved and enhanced signal processing method implemented in the capnography system. To give a measure of reliability to the calculated value, it is necessary to determine possible calculation errors and identify their potential impact on the results (Khurram et al., 2022). The uncertainties associated with the results of signal processing analysis can be expressed as the intervals at which error values are likely to be present. The process of systematically quantifying error estimates is called uncertainty analysis. There are two types of uncertainty analysis practiced as per the standard, which are random uncertainty and systematic uncertainty. Random uncertainty that can be evaluated by repeating calculations and statistical spread of results. Digital signal processing can be repeated using the same raw data. High accuracy is achieved if the calculations are repeatable and consistent if there is little variation in results (Albaba et al., 2021). Systematic uncertainty represents a potential systematic difference in the analysis. This may occur due to an incorrect formula in the calculation or a programming error. Uncertainty in this category cannot be evaluated by iterative calculations and must be evaluated considering potential factors that affect the accuracy of the calculation (Gómez–Echavarría et al., 2020; Terrien et al., 2010). RGM standards and signal processing quality assurance manual have denoted that the quality and precision of signal processing of RGM tool, as capnography must be complying with the control of computer system, testing of signal processing software, monitoring processes and the feedback loops (Oliveri et al., 2019). Capnography is nonlinear quasiperiodic waveform, which reflects the flow of CO_2 varies. Capnography with multiple applications in respiratory and pulmonary assessments requires appropriate set of criteria and aspects to provide accurate result, regardless of the predominant noises appearing in the system. Capnography waveform encounters noises arising from bandwidth and high-frequency components (Chen & Cai, 2019; Mooney et al., 2008). The effect of the noises will be seen in the capnogram waveform, which results in uneven and blunt end curves. Hence, signal filtering is a critical point in the capnography system, to deliver a sharp and accurate waveform. The standard has defined that filtering is a denoising process, which estimates a clean capnogram signal from its noisy state. The denoiser target to deliver clean signal using a denoising method, known as signal filtering techniques. There are several signal filtering techniques are applied in denoising the capnogram signals (Butusov et al., 2018). Low-pass filtering is one of common type of denoiser used for high-frequency operating systems. Several systems demand slow dynamics, where the low–pass filters are implemented. The filter then reduces the signal amplifying trend and delivers smooth and palatable signals. Conversely, high-pass filter is a common denoising technique used in systems, which requires shift from slow drift to high drift signal of interest. The filter removes slow noises prior to the requirement of data analysis. It is a standard tool to remove DC components to enable satisfying implementation of limited range digital system (Ardoint & Lorenzi, 2010; Baxevanaki et al., 2019). Besides that, band-pass filtering is a traditional way of data analysis, which isolates a particular band and applies the filter. Band-pass filter interprets capnogram signal based on the frequency bands. Capnogram signal processing usually applies more than one band–pass filter for delivering smooth capnogram waveform (Chatterjee et al., 2020). Fixed–coefficient filter is traditionally applied in capnography to suppress the artifacts caused by lung and chest compression. The filter suppresses the spectral content of capnogram waveform above 1 Hz and developed a digital impulse response. Adaptive filtering is another type of denoiser, which is implemented in capnogram when other types of filters not able to withstand the artifacts. This filter adapts to the artifact and alters the parameter of the adaptive filters (Kumar et al., 2021). Filters are ubiquitous in spectrogram studies and an important part to eliminate the noise and reveal targeted signal. Implementing one or more filter in a capnography system is obligatory, yet the potential effect into system is critical to be considered for developing novel capnography for asthma monitoring (de Cheveigné & Nelken, 2019). Fig. 7.2 illustrates the flow of applying more than one filter in a system for displaying smooth signals.

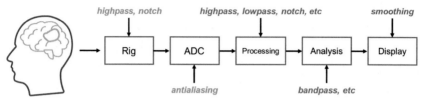

Figure 7.2 Anatomy of several filters implementation in a system.

7.4 Available technology for the development of CO$_2$ measurement device

Fig. 7.3 depicts the development of a CO_2 measurement device as a function of time or volume using side- and main-stream technologies. A CO_2 sensor is positioned inside the main unit, away from the subject, to detect CO_2 molecules in the sidestream approach. The sample is aspirated from the sampling tube by a small pump to the CO_2 sensor, ensuring that the sidestream capnograph is reliable for both adults and children (Fabius et al., 2016; Rasera et al., 2015). Meanwhile, the mainstream technology holds the CO_2 sensor away from the gas sampling site and the sensing mechanism takes place through a data transmission unit. In the main-stream approach, the CO_2 sensor is positioned between the endotracheal tube and the breathing circuit. As a result, no sampling tube, pump, motor, or scavenges are involved (Brown et al., 1998; Jaffe & Orr, 2010; Maestri et al., 2013). Additionally, the capnography signal show faster response time, a simpler mechanism, and a more accurate sample rate. The CO_2 sensors for mainstream set-up are expensive and hefty, requiring solid-state

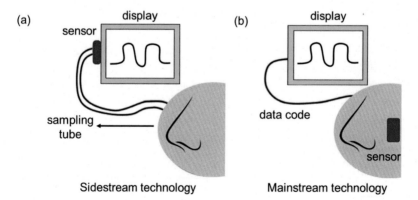

Figure 7.3 CO_2 measurement technology: (A) side-stream and (B) main-stream.

sources, improved optics, and downsizing to ensure the steady gas sampling system (Nurifhan et al., 2016). As a result, when compared to the mainstream technique, the sidestream technique is found to be more convenient, easy, and has no concerns with sterilization. It can also be employed when a patient is sitting in comfortable position (Brown et al., 2013). In this chapter, design of a sidestream real time CO_2 measurement device is explained. The lightweight, portable, and inexpensive capnography tool developed is highly recommended for use by medical personnel at clinical setting and people at home environment (Kesten et al., 1990; Lamba et al., 2009).

7.5 Design of CO_2 measurement tool: an overview

This chapter encloses the design of capnography tool and the system on the principle of operation, in relation to the sensors, electronic components and algorithms used in the designed study. Fig. 7.4 shows the illustration of patient monitored for asthma using designed capnography tool and the digital image of the tool.

The designed CO_2 measurement tool comprises of four parts, which are CO_2 acquisitions unit, processing unit, real-time control (RTC), and a display unit. Sampling tube collects CO_2 gas from subject and transfers to the microcontroller for computational and transmission activities (Brown

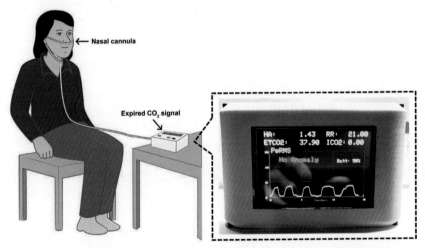

Figure 7.4 Representation of patient fixed with nasal cannula to monitor asthmatic condition. The digital image shows prototype design of developed capnography tool for real-time CO_2 measurement.

et al., 1998; Mieloszyk et al., 2014). Moreover, serial communication enables CO_2 signal extraction and its display on thin film transistor (TFT). Based on the CO_2 measurement and analyses workflow, it is evident the design and computations are far inexpensive compared to existing methods (Kean & Malarvili, 2009). A real-time control unit is used for data logging into SD card. The process is controlled by processing unit. Fig. 7.5 shows designed capnography tool block diagram.

7.5.1 CO_2 acquisition unit

CO_2 sensor is the most critical component, as it effectively detects the CO_2 gas from sampling tube and display the accurate capnogram waveform. In comparison to several NDIR CO_2 sensors, COMET CO_2 sensor has been recognized as a promising sensor to be used in capnography (Manifold et al., 2013). Table 7.3 shows the existing CO_2 sensors and the high-performance specification owned by COMET sensor. These specifications indicate that the COMET CO_2 sensor is highly selective and sensitive toward CO_2 gas compared to other sensors (Domingo et al., 2010). In addition, the sampling rate (100, samples per second-sps) of the COMET is comparatively high that provides more precise and detailed analysis of the CO_2 signal. Hence, the designed capnography tool uses NDIR CO_2 sensor equivalent to COMET to acquire the CO_2 signal. NDIR sensor absorbs the CO_2 molecules at a specific wavelength (4.3 µm) in the infrared region and follows the Beer–Lambert law (Howe et al., 2011). With this, the sensor able to differentiate CO_2 gas with other gases such as O_2, N_2O and water vapor, which are present in human breath. In the effort for a novel

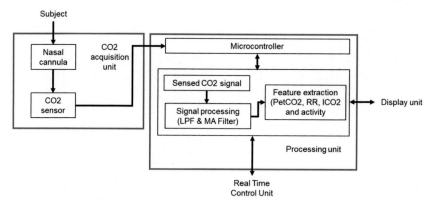

Figure 7.5 Flow diagram of designed capnography tool for real-time CO_2 measurement.

Table 7.3 Comparison of specifications of various NDIR CO_2 sensors used in development of CO_2 measurement device.

CO_2 sensor	Warm-up time (s)	Response time (s)	Weight (g)	Refs.
Sprint IR	<60	0.05	8	Bautista et al. (2013), Nurifhan et al. (2016)
MG811	300	<60	60	Zaharudin et al. (2014)
COZIR	<10	4–120	20	Gibson and MacGregor, (2013)
COMET	2–15	0.028	<7.0	Data et al. (2015)
MH410	90	<30	20	Winsen, (2016)

innovation, a real-time CO_2 measurement device is developed using a first order low-pass filer and 100 Hz cut off frequency (Thompson & Jaffe, 2005). The chapter presents the details on the developed device components, feature computation, and algorithm transmission for determining asthma severity based on the displaying the CO_2 extracted features. The presented work provides envision on the feature retrieval, filtration of signal and tuning the performance of capnography device. Then, the CO_2 gas from subject is passed to the CO_2 sensor through a sampling tube. The sampling tube (7 feet long) is connected to the water trap and to the subject to transport the CO_2 gas sample (Ffarcs & Da, 1992). The built-in water trap is extremely sensitive, trapping moisture and other secretions while keeping the CO_2 signal in firm placement. The 0.2 μ hydrophobic and particulate (200 μ) filters are combined with a water trap to remove any residual water vapor from the gas sample while maintaining a laminar flow to reduce CO_2 waveform distortion. All above-mentioned features provided a solution to the existing drawbacks in capnography at clinical utilization (Yaron et al., 1996). Thereafter, CO_2 samples are drawn from the sampling tube to the CO_2 sensor at the speed of 50 mL/min using a pump and DC motor that enable use with the developed device at a higher respiratory rate (150 bpm). Further, CO_2 samples are drawn with the sampling rate of 100 Hz through an infrared CO_2 sensor. The sensor has 3-pin Universal Asynchronous Receiver/Transmitter (UART) communication at the transistor-transistor logic (TTL) level for transmitting and receiving the CO_2 data from the sensor to the microcontroller and vice-versa.

7.5.2 Processing unit

Microcontroller is the key component for processing CO_2 gas samples after the CO_2 sensor. In the presented design, Arduino Mega 2560 microcontroller is used to process the CO_2 gas samples for computation and transmission. The Arduino (Guthrie et al., 2007) is a free and open-source microcontroller board based on the Atmega2560. It is used in the same way that UARTs are to read TTL (5V) inputs from buttons, external switches, sensors, displays, and many others (Corbo et al., 2005). Arduino consists of 16 MHz crystal oscillator, a USB connection, a power jack, an ICSP header, and a reset button, which are required for driving the microcontroller (Nagurka et al., 2014). It can drive by connecting a USB cable to a computer or power with an AC to DC adapter or battery (Kassabian et al., 1982). The Arduino board, also, can operate on an external power supply of 6−20 V. However, the recommended power range is 7—12 V. Besides, it displays the CO_2 data from a CO_2 sensor on the serial port simultaneously, which is not possible with Arduino Uno for higher baud rate (19,200, 115,200).

7.5.3 Display unit

The Adafruit TFT is used for displaying the CO_2 signal and other features. It can drive with minimum power supply (3 V). It has higher resolution (240 × 320 pixels) than organic light emitting diode (OLED − 128 × 64). It also has an inbuilt controller with RAM buffering, which turn down the controlling work (Nik Hisamuddin et al., 2009). There are two modes for the display to be integrated with microcontroller. First is 8-bit interface to read and write the display using 8 and 5, digital data lines and control digital lines, respectively. Second mode is SPI, which has five pins for data configuration. In comparison, 8-bit mode shows high speed data transmission than SPI. Conversely, SPI is flexible and simple to be integrated with multiple microcontrollers (Azab et al., 2015). Moreover, SPI able to fit with microSD card, whereas 8-bit mode is unable for the purpose. Therefore, the TFT-2.8″ display mode is used in the study to develop simple and easy for use capnography tool.

7.5.4 Real time control unit

Real-time control module (DS3231) enables the data storage of subject information, according to the date and time of study. A real-time clock and temperature-compensated crystal oscillator integrated control module is highly accurate with low-cost. As an alternative to the main power, the module is integrated with lithium battery (3 V) to keep maintain the accurate data set. The control module able to trace accurate time from seconds to year, including the change of days occurs in leap year (Howe et al., 2011). Real-time control module is simple and able for easy integration with microcontroller.

7.6 Real time human respiration CO₂ measurement device

This chapter presents the design of a novel capnography tool as the light weight and low-cost human respiration CO_2 measurement prototype with a high potential use at clinical settings and home environment. A preliminary model of the proposed set-up is presented in Fig. 7.6. Two PCB boards are used for the proposed complete prototype device. First PCB

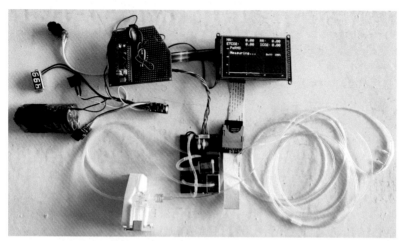

Figure 7.6 Internal arrangement of components in the prototype design of capnography tool.

carries CO_2 sensor, pump, DC motor, and a battery that constantly provides a power supply of 5 V to the sensor. The second PCB accommodates microcontroller and RTC unit, whereas TFT is fixed at the top of this board. There are three simple mechanical switches, of which two are installed to offer the consistent power to the microcontroller and sensor and the last one is employed to control the recording of CO_2 data and its features into SD card, which is placed into SD card slot of the TFT. Moreover, LED records the data and recognizes the fault found in the SD card. SD card undergoes formatting, when the switch is pressed, and LED turns off. This feature makes the measurement tool easier to operate, and user-friendly. The CO_2 acquisition unit, which consists of a sensor, motor, pump, and a 5 V battery, is located on level one, while the processing and RTC unit are located on level two. Both levels are housed in a case measuring 12.40 × 14 × 7.6 cm, with a top cover created and produced using Solid Works (version) and a 3-D printer, respectively. As a result, the device's weight is reduced by about 500 g. To enable appropriate access to the user interface, all the components are positioned close to each other on the top cover of the box.

Fig. 7.7 denotes the CO_2 waveform and associated features as the sample line is fastened. CO_2 signals are recorded for approximately 2 minutes from two different healthy subject using proposed capnograph tool at the sampling rate of 100 Hz. As discussed in Chapter 5, a low pass filter (fc, 10 Hz) and moving average filter (span, 8) are integrated with the device

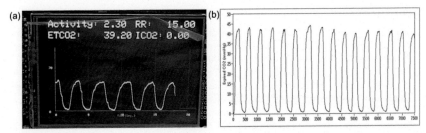

Figure 7.7 (A) Digital image of CO_2 signal acquisition from developed capnography tool and (B) capnogram waveform at 100 Hz.

that provides a smooth signal and Pearson correlation coefficients is calculated for each subject before and after the filter to choose the optimum span, to avoid the loss of data. The obtained parameters and CO_2 signal are saved into SD card (8 GB) through an RTC unit (DS3231). Each data point is saved inside a folder with respect to date, day, and time with the delayed of 10 ms that makes device simple to pursue the subject's information in the future. The code for each folder has a designated ID containing the current date. The saving procedure is controlled by a mechanical switch that facilities to save the data. After every press, a new file is generated according to hours, minutes, and second of the day that depends upon the need of the subjects or observer.

Hereby, this chapter denotes the design of simple and straightforward capnography tool, which are incorporated with high performance NDIR CO_2 sensor, inexpensive electronic components and precisely designed algorithms for feature extraction and transmissions. The capnography tool is designed as it is suitable for real-time CO_2 measurement at medical facilities and home-environment. With the regard, the design of capnography tool is easy to be operated by people, with easily understandable operating procedures. Moreover, the novel application of the device at home —environment is expected to ease people to identify individual respiratory condition, especially asthma. A clear and detailed storage of data recorded in the SD card further ease the medical personnel to monitor the asthmatic condition of a patient at current and previous conditions. With an alternative and innovative approach performed for the simple and straightforward design and system, the real-time CO_2 measuring capnography tool discussed in this chapter is recommended for application in diagnosing respiratory and cardiopulmonary illnesses, to save millions of fatal deaths from the noncommunicable diseases.

References

Agrahari, S., & Singh, A. K. (2021). Concept drift detection in data stream mining: A literature review. *Journal of King Saud University — Computer and Information Sciences, xxxx*. https://doi.org/10.1016/j.jksuci.2021.11.006

Albaba, A., Castro, I., Borzée, P., Buyse, B., Testelmans, D., Varon, C., Van Huffel, S., & Torfs, T. (2021). Automatic quality assessment of capacitively-coupled bioimpedance signals for respiratory activity monitoring. *Biomedical Signal Processing and Control, 68*(February). https://doi.org/10.1016/j.bspc.2021.102775

de Andrade, F. M., Sales, R., da Silva, N. C., & Pimentel, M. F. (2022). Calibration with virtual standards for monitoring biodiesel production using a miniature NIR spectrometer. *Talanta, 243*(November 2021). https://doi.org/10.1016/j.talanta.2022.123329

de Andrade e Silva, A. G., Gomes, H. M., & Batista, L. V. (2022). A collaborative deep multi-task learning network for face image compliance to ISO/IEC 19794-5 standard. *Expert Systems with Applications, 198*(February), 116756. https://doi.org/10.1016/j.eswa.2022.116756

An, K., Li, S., Shao, S., Liu, P., Liu, J., Chen, L., Wei, J., Zheng, Y., Liu, Q., & Li, C. (2022). Evaluation of the fracture strength of ultra-thick diamond plate by the three-point bending ISO standard method. *Ceramics International, 48*(12), 17942—17949. https://doi.org/10.1016/j.ceramint.2022.03.101

Ardoint, M., & Lorenzi, C. (2010). Effects of lowpass and highpass filtering on the intelligibility of speech based on temporal fine structure or envelope cues. *Hearing Research, 260*(1—2), 89—95. https://doi.org/10.1016/j.heares.2009.12.002

Azab, N. Y., El Mahalawy, I. I., Abd El Aal, G. A., & Taha, M. H. (2015). Breathing pattern in asthmatic patients during exercise. *Egyptian Journal of Chest Diseases and Tuberculosis, 64*(3), 521—527. https://doi.org/10.1016/j.ejcdt.2015.02.009

Bautista, C., Patel, B., Shah, M., Connie, L., Seifer, B., & Facas, G. (2017). Portable capnography.

Baxevanaki, K., Kapoulea, S., Psychalinos, C., & Elwakil, A. S. (2019). Electronically tunable fractional-order highpass filter for phantom electroencephalographic system model implementation. *AEU — International Journal of Electronics and Communications, 110*, 152850. https://doi.org/10.1016/j.aeue.2019.152850

Benedetto, F., & Tosti, F. (2017). A signal processing methodology for assessing the performance of ASTM standard test methods for GPR systems. *Signal Processing, 132*, 327—337. https://doi.org/10.1016/j.sigpro.2016.06.030

Boehler, C. N., Appelbaum, L. G., Krebs, R. M., Hopf, J. M., & Woldorff, M. G. (2012). The influence of different stop-signal response time estimation procedures on behavior-behavior and brain-behavior correlations. *Behavioural Brain Research, 229*(1), 123—130. https://doi.org/10.1016/j.bbr.2012.01.003

Brown, R. H., Brooker, A., Wise, R. A., Reynolds, C., Loccioni, C., Russo, A., & Risby, T. H. (2013). Forced expiratory capnography and chronic obstructive pulmonary disease (COPD). *Journal of Breath Research, 7*(1). https://doi.org/10.1088/1752-7155/7/1/017108

Brown, L. H., Gough, J. E., & Seim, R. H. (1998). Can quantitative capnometry differentiate between cardiac and obstructive causes of respiratory distress? *Chest, 113*(2), 323—326. https://doi.org/10.1378/chest.113.2.323

Butusov, D., Karimov, T., Voznesenskiy, A., Kaplun, D., Andreev, V., & Ostrovskii, V. (2018). Filtering techniques for chaotic signal processing. *Electronics (Switzerland), 7*(12). https://doi.org/10.3390/electronics7120450

de Carellán Mateo, A. G., & Casamián-Sorrosal, D. (2022). The perioperative management of small animals with previously implanted pacemakers undergoing anaesthesia. *Veterinary Anaesthesia and Analgesia, 49*(1), 7—17. https://doi.org/10.1016/j.vaa.2021.05.007

Castillo-Martinez, A., Medina-Merodio, J. A., Gutierrez-Martinez, J. M., & Fernández-Sanz, L. (2021). Proposal for a maintenance management system in industrial environments based on ISO 9001 and ISO 14001 standards. *Computer Standards and Interfaces, 73.* https://doi.org/10.1016/j.csi.2020.103453

Charlton, P. H., Bonnici, T., Tarassenko, L., Clifton, D. A., Beale, R., Watkinson, P. J., & Alastruey, J. (2021). An impedance pneumography signal quality index: Design, assessment and application to respiratory rate monitoring. *Biomedical Signal Processing and Control, 65*(October 2020), 102339. https://doi.org/10.1016/j.bspc.2020.102339

Chatterjee, S., Thakur, R. S., Yadav, R. N., Gupta, L., & Raghuvanshi, D. K. (2020). Review of noise removal techniques in ECG signals. *IET Signal Processing, 14*(9), 569—590. https://doi.org/10.1049/iet-spr.2020.0104

Chee, R. C., Ahmad, R., Zakaria, M. I., & Yahya, M. F. (2021). The assessment of end-tidal capnography waveform interpretation and its clinical application for emergency residents in Malaysia: A cross-sectional study. *Eurasian Journal of Emergency Medicine, 20*(3), 161—171. https://doi.org/10.4274/eajem.galenos.2021.83097

Chen, J., & Cai, Z. (2019). Cardinal MK-spline signal processing: Spatial interpolation and frequency domain filtering. *Information Sciences, 495*(2016), 116—135. https://doi.org/10.1016/j.ins.2019.04.056

de Cheveigné, A., & Nelken, I. (2019). Filters: When, why, and how (not) to use them. *Neuron, 102*(2), 280—293. https://doi.org/10.1016/j.neuron.2019.02.039

Chin, S. T., Romano, A., Doran, S. L. F., & Hanna, G. B. (2018). Cross-platform mass spectrometry annotation in breathomics of oesophageal-gastric cancer. *Scientific Reports, 8*(1), 1—10. https://doi.org/10.1038/s41598-018-22890-w

Corbo, J., Bijur, P., Lahn, M., & Gallagher, E. J. (2005). Concordance between capnography and arterial blood gas measurements of carbon dioxide in acute asthma. *Annals of Emergency Medicine, 46*(4), 323—327. https://doi.org/10.1016/j.annemergmed.2004.12.005

Data, R.U.S.A., Falligant, C., Klaus, J., Grove, C., Montgomery, F. J., & Toombs, C. (2015). *(12) Patent Application Publication (10) Pub. No.: US 2015/0032019 A1. 1*(19).

De Battista, H., Picó, J., Picó-Marco, E., & Vignoni, A. (2019). Biomolecular signal tracker with fast time response. *IFAC-PapersOnLine, 52*(26), 1—6. https://doi.org/10.1016/j.ifacol.2019.12.227

Deacon, A. J., & Pratt, O. W. (2021). Measurement of pulse oximetry, capnography and pH. *Anaesthesia and Intensive Care Medicine, 22*(3), 185—189. https://doi.org/10.1016/j.mpaic.2021.01.005

Dinh, T. V., Choi, I. Y., Son, Y. S., & Kim, J. C. (2016). A review on non-dispersive infrared gas sensors: Improvement of sensor detection limit and interference correction. *Sensors and Actuators, B: Chemical, 231*, 529—538. https://doi.org/10.1016/j.snb.2016.03.040

Djerdj, T., Peršić, V., Hackenberger, K., Hackenberger, D. K., & Hackenberger, B. K. (2021). A low-cost versatile system for continuous real-time respiratory activity measurement as a tool in environmental research. *Measurement: Journal of the International Measurement Confederation, 184*(July). https://doi.org/10.1016/j.measurement.2021.109928

Domingo, C., Blanch, L., Murias, G., & Luján, M. (2010). State-of-the-art sensor technology in Spain: Invasive and non-invasive techniques for monitoring respiratory variables. *Sensors, 10*(5), 4655—4674. https://doi.org/10.3390/s100504655

Duncan, A., & Pratt, O. W. (2021). Measurement of gas concentrations. *Anaesthesia and Intensive Care Medicine, 22*(3), 190—193. https://doi.org/10.1016/j.mpaic.2021.01.009

Fabius, T. M., Eijsvogel, M. M., Van Der Lee, I., Brusse-Keizer, M. G. J., & De Jongh, F. H. (2016). Volumetric capnography in the exclusion of pulmonary embolism at the emergency department: A pilot study. *Journal of Breath Research, 10*(4). https://doi.org/10.1088/1752-7163/10/4/046016

Ffarcs, H. S. L. M., & Da, Y. D. (1992). Equipment inspiratory valve malfunction in a circle system: Pitfalls in capnography. *Anaesthesia*, 997—999.

Fleming, L., Gibson, D., Song, S., Li, C., & Reid, S. (2018). Reducing N_2O induced cross-talk in a NDIR CO_2 gas sensor for breath analysis using multilayer thin film optical interference coatings. *Surface and Coatings Technology, 336*, 9—16. https://doi.org/10.1016/j.surfcoat.2017.09.033

Fuss, F. K., Tan, A. M., & Weizman, Y. (2019). 'Electrical viscosity' of piezoresistive sensors: Novel signal processing method, assessment of manufacturing quality, and proposal of an industrial standard. *Biosensors and Bioelectronics, 141*(June), 111408. https://doi.org/10.1016/j.bios.2019.111408

Gibson, D., & MacGregor, C. (2013). A novel solid state non-dispersive infrared CO_2 gas sensor compatible with wireless and portable deployment. *Sensors (Switzerland), 13*(6), 7079—7103. https://doi.org/10.3390/s130607079

Gómez-Echavarría, A., Ugarte, J. P., & Tobón, C. (2020). The fractional fourier transform as a biomedical signal and image processing tool: A review. *Biocybernetics and Biomedical Engineering, 40*(3), 1081—1093. https://doi.org/10.1016/j.bbe.2020.05.004

Guthrie, B. D., Adler, M. D., & Powell, E. C. (2007). End-tidal carbon dioxide measurements in children with acute asthma. *Academic Emergency Medicine, 14*(12), 1135—1140. https://doi.org/10.1197/j.aem.2007.08.007

He, H., Guo, J., Illés, B., Géczy, A., Istók, B., Hliva, V., Török, D., Kovács, J. G., Harmati, I., & Molnár, K. (2021). Monitoring multi-respiratory indices via a smart nano-fibrous mask filter based on a triboelectric nanogenerator. *Nano Energy, 89*. https://doi.org/10.1016/j.nanoen.2021.106418

He, X., Nie, B., Chen, W., Wang, E., Dou, L., Wang, Y., Liu, M., & Hani, M. (2012). Research progress on electromagnetic radiation in gas-containing coal and rock fracture and its applications. *Safety Science, 50*(4), 728—735. https://doi.org/10.1016/j.ssci.2011.08.044

Hosseini Teshnizi, S., Hayavi Haghighi, M. H., & Alipour, J. (2021). Evaluation of health information systems with ISO 9241-10 standard: A systematic review and meta-analysis. *Informatics in Medicine Unlocked, 25*, 100639. https://doi.org/10.1016/j.imu.2021.100639

Howe, T. A., Jaalam, K., Ahmad, R., Sheng, C. K., & Nik Ab Rahman, N. H. (2011). The use of end-tidal capnography to monitor non-intubated patients presenting with acute exacerbation of asthma in the emergency department. *Journal of Emergency Medicine, 41*(6), 581—589. https://doi.org/10.1016/j.jemermed.2008.10.017

Huang, Z., & Wang, M. (2021). A review of electroencephalogram signal processing methods for brain-controlled robots. *Cognitive Robotics, 1*(May), 111—124. https://doi.org/10.1016/j.cogr.2021.07.001

Ibrahim, M. F., Hod, R., Ahmad Tajudin, M. A. B., Wan Mahiyuddin, W. R., Mohammed Nawi, A., & Sahani, M. (2022). Children's exposure to air pollution in a natural gas industrial area and their risk of hospital admission for respiratory diseases. *Environmental Research, 210*(December 2021), 112966. https://doi.org/10.1016/j.envres.2022.112966

Jabłoński, I., Polak, A. G., & Mroczka, J. (2011). Preliminary study on the accuracy of respiratory input impedance measurement using the interrupter technique. *Computer Methods and Programs in Biomedicine, 101*(2), 115—125. https://doi.org/10.1016/j.cmpb.2010.11.003

Jaffe, M. B., & Orr, J. (2010). Continuous monitoring of respiratory flow and CO_2. *IEEE Engineering in Medicine and Biology Magazine, 29*(2), 44—52. https://doi.org/10.1109/memb.2009.935712

Jashnsaz, H., Fox, Z. R., Munsky, B., & Neuert, G. (2021). Building predictive signaling models by perturbing yeast cells with time-varying stimulations resulting in distinct signaling responses. *STAR Protocols, 2*(3), 100660. https://doi.org/10.1016/j.xpro.2021.100660

Johri, S., & Kumar, D. (2021). Evaluation of effect of ISO 9001:2008 standard implementation on TQM Parameters in manufacturing & production processes performance in small Enterprises. *Materials Today: Proceedings, xxxx.* https://doi.org/10.1016/j.matpr.2020.12.696

Kassabian, J., Miller, K. D., & Lavietes, M. H. (1982). Respiratory center output and ventilatory timing in patients with acute airway (asthma) and alveolar (pneumonia) disease. *Chest, 81*(5), 536–543. https://doi.org/10.1378/chest.81.5.536

Kean, T. T., & Malarvili, M. B. (2009). Analysis of capnography for asthmatic patient. ICSIPA09 – 2009 IEEE International Conference on signal and image processing applications. *Conference Proceedings,* 464–467. https://doi.org/10.1109/ICSIPA.2009.5478699

Keijzer, I. N., & Scheeren, T. W. L. (2021). Perioperative hemodynamic monitoring: An overview of current methods. *Anesthesiology Clinics, 39*(3), 441–456. https://doi.org/10.1016/j.anclin.2021.03.007

Kesten, S., Maleki-Yazdi, R., Sanders, B. R., Wells, J. A., McKillop, S. L., Chapman, K. R., & Rebuck, A. S. (1990). Respiratory rate during acute asthma. *Chest, 97*(1), 58–62. https://doi.org/10.1378/chest.97.1.58

Khurram, O. U., Gransee, H. M., Sieck, G. C., & Mantilla, C. B. (2022). Automated evaluation of respiratory signals to provide insight into respiratory drive. *Respiratory Physiology and Neurobiology, 300*(February), 103872. https://doi.org/10.1016/j.resp.2022.103872

Kumar, A., Tomar, H., Mehla, V. K., Komagiri, R., & Kumar, M. (2021). Stationary wavelet transform based ECG signal denoising method. *ISA Transactions, 114,* 251–262. https://doi.org/10.1016/j.isatra.2020.12.029

Lamba, S., Gluckman, W., Nagurka, R., Rosania, A., Bechmann, S., Langley, D. J., Scott, S., & Compton, S. (2009). Initial out-of-hospital end-tidal carbon dioxide measurements in adult asthmatic patients. *Annals of Emergency Medicine, 54*(3), S51. https://doi.org/10.1016/j.annemergmed.2009.06.193

Lermuzeaux, M., Meric, H., Sauneuf, B., Girard, S., Normand, H., Lofaso, F., & Terzi, N. (2016). Superiority of transcutaneous CO_2 over end-tidal CO_2 measurement for monitoring respiratory failure in nonintubated patients: A pilot study. *Journal of Critical Care, 31*(1), 150–156. https://doi.org/10.1016/j.jcrc.2015.09.014

Lewis, P. M., Vahsen, S. E., Seong, I. S., Hedges, M. T., Jaegle, I., & Thorpe, T. N. (2015). Absolute position measurement in a gas time projection chamber via transverse diffusion of drift charge. *Nuclear Instruments and Methods in Physics Research, Section A: Accelerators, Spectrometers, Detectors and Associated Equipment, 789,* 81–85. https://doi.org/10.1016/j.nima.2015.03.024

Li, B., Feng, C., Wu, H., Jia, S., & Dong, L. (2022). Calibration-free mid-infrared exhaled breath sensor based on BF-QEPAS for real-time ammonia measurements at ppb level. *Sensors and Actuators B: Chemical, 358*(December 2021), 131510. https://doi.org/10.1016/j.snb.2022.131510

Li, Y., Jiang, R., Tapia, J., Wang, S., & Sun, W. (2020). Structural damage identification based on short-time temporal coherence using free-vibration response signals. *Measurement: Journal of the International Measurement Confederation, 151,* 107209. https://doi.org/10.1016/j.measurement.2019.107209

Lin, K. C., Yang, J. W., Ho, P. Y., Yen, C. Z., Huang, H. W., Lin, H. Y., Chung, J., & Chen, G. Y. (2022). Development of an alveolar chip model to mimic respiratory conditions due to fine particulate matter exposure. *Applied Materials Today, 26,* 101281. https://doi.org/10.1016/j.apmt.2021.101281

Maestri, R., Bruschi, C., Olmetti, F., La Rovere, M. T., & Pinna, G. D. (2013). Assessment of the peripheral ventilatory response to CO_2 in heart failure patients: Reliability of the single-breath test. *Physiological Measurement, 34*(9), 1123–1132. https://doi.org/10.1088/0967-3334/34/9/1123

Maho, P., Herrier, C., Livache, T., Comon, P., & Barthelmé, S. (2022). A calibrant-free drift compensation method for gas sensor arrays. *Chemometrics and Intelligent Laboratory Systems, 225*(April), 104549. https://doi.org/10.1016/j.chemolab.2022.104549

Manifold, C. A., Davids, N., Villers, L. C., & Wampler, D. A. (2013). Capnography for the nonintubated patient in the emergency setting. *Journal of Emergency Medicine, 45*(4), 626–632. https://doi.org/10.1016/j.jemermed.2013.05.012

Martini, N., Koukou, V., Fountos, G., Valais, I., Bakas, A., Ninos, K., Kandarakis, I., Panayiotakis, G., & Michail, C. (2018). Towards the enhancement of medical imaging with non-destructive testing (NDT) CMOS sensors. Evaluation following IEC 62220-1-1:2015 international standard. *Procedia Structural Integrity, 10*, 326–332. https://doi.org/10.1016/j.prostr.2018.09.045

Mettelman, R. C., Allen, E. K., & Thomas, P. G. (2022). Mucosal immune responses to infection and vaccination in the respiratory tract. *Immunity, 55*(5), 749–780. https://doi.org/10.1016/j.immuni.2022.04.013

Mieloszyk, R. J., Verghese, G. C., Deitch, K., Cooney, B., Khalid, A., Mirre-González, M. A., Heldt, T., & Krauss, B. S. (2014). Automated quantitative analysis of capnogram shape for COPD-normal and COPD-CHF classification. *IEEE Transactions on Biomedical Engineering, 61*(12), 2882–2890. https://doi.org/10.1109/TBME.2014.2332954

Mimoz, O., Benard, T., Gaucher, A., Frasca, D., & Debaene, B. (2012). Accuracy of respiratory rate monitoring using a non-invasive acoustic method after general anaesthesia. *British Journal of Anaesthesia, 108*(5), 872–875. https://doi.org/10.1093/bja/aer510

Mooney, A., Keating, J. G., & Pitas, I. (2008). A comparative study of chaotic and white noise signals in digital watermarking. *Chaos, Solitons and Fractals, 35*(5), 913–921. https://doi.org/10.1016/j.chaos.2006.05.073

Nagurka, R., Bechmann, S., Gluckman, W., Scott, S. R., Compton, S., & Lamba, S. (2014). Utility of initial prehospital end-tidal carbon dioxide measurements to predict poor outcomes in adult asthmatic patients. *Prehospital Emergency Care, 18*(2), 180–184. https://doi.org/10.3109/10903127.2013.851306

Narimani, M., Narimani, T., & Alizadeh, R. (2022). Measuring serum sodium levels using blood gas analyzer and auto analyzer in heart and lung disease patients: A cross-sectional study. *Annals of Medicine and Surgery, 78*(April), 103713. https://doi.org/10.1016/j.amsu.2022.103713

Nguyen, D. C. T., Dowling, J., Ryan, R., McLoughlin, P., & Fitzhenry, L. (2021). Pharmaceutical-loaded contact lenses as an ocular drug delivery system: A review of critical lens characterization methodologies with reference to ISO standards. *Contact Lens and Anterior Eye, 44*(6), 101487. https://doi.org/10.1016/j.clae.2021.101487

Nik Hisamuddin, N. A. R., Rashidi, A., Chew, K. S., Kamaruddin, J., Idzwan, Z., & Teo, A. H. (2009). Correlations between capnographic waveforms and peak flow meter measurement in emergency department management of asthma. *International Journal of Emergency Medicine, 2*(2), 83–89. https://doi.org/10.1007/s12245-009-0088-9

Ni, J., Zhang, Y., Ding, F., Zhan, X. S., & Hayat, T. (2021). Parameter estimation algorithms of linear systems with time-delays based on the frequency responses and harmonic balances under the multi-frequency sinusoidal signal excitation. *Signal Processing, 181*, 107904. https://doi.org/10.1016/j.sigpro.2020.107904

Nurifhan, A., Krishnan, M. B., Prakash, O., & Ahmed, S. (2016). Portable respiratory CO_2 monitoring device for early screening of asthma. *Fourth International Conference on*

Advances in Computing, Electronics and Communication, 12, 90—94. https://doi.org/10.15224/978-1-63248-113-9-61

Nurifhan, A., Prakash, O., & Ahmed, S. (2016). *Portable respiratory CO_2 monitoring device for early screening of asthma.* https://doi.org/10.15224/978-1-63248-113-9-61. December, 90—94.

Olarte, O., De Keyser, R., & Ionescu, C. M. (2016). Fan-based device for non-invasive measurement of respiratory impedance: Identification, calibration and analysis. *Biomedical Signal Processing and Control, 30,* 127—133. https://doi.org/10.1016/j.bspc.2016.06.004

Oliveri, P., Malegori, C., Simonetti, R., & Casale, M. (2019). The impact of signal preprocessing on the final interpretation of analytical outcomes — A tutorial. *Analytica Chimica Acta, 1058,* 9—17. https://doi.org/10.1016/j.aca.2018.10.055

Onyancha, R. B., Ukhurebor, K. E., Aigbe, U. O., Osibote, O. A., Kusuma, H. S., Darmokoesoemo, H., & Balogun, V. A. (2021). A systematic review on the detection and monitoring of toxic gases using carbon nanotube-based biosensors. *Sensing and Bio-Sensing Research, 34,* 100463. https://doi.org/10.1016/j.sbsr.2021.100463

Paniccia, A., Rozner, M., Jones, E. L., Townsend, N. T., Varosy, P. D., Dunning, J. E., Girard, G., Weyer, C., Stiegmann, G. V., & Robinson, T. N. (2014). Electromagnetic interference caused by common surgical energy-based devices on an implanted cardiac defibrillator. *American Journal of Surgery, 208*(6), 932—936. https://doi.org/10.1016/j.amjsurg.2014.09.011

Pella, L., Lambert, C., McArthur, B., West, C., Hernandez, M., Green, K., Sousa, M., Brast, S., & Long, M. (2018). Systematic review to develop the clinical practice guideline for the use of capnography during procedural sedation in radiology and imaging settings: A report of the association for radiologic & imaging nursing capnography task force. *Journal of Radiology Nursing, 37*(3), 163—172. https://doi.org/10.1016/j.jradnu.2018.07.003

Rasera, C. C., Gewehr, P. M., & Domingues, A. M. T. (2015). PETCO$_2$ measurement and feature extraction of capnogram signals for extubation outcomes from mechanical ventilation. *Physiological Measurement, 36*(2), 231—242. https://doi.org/10.1088/0967-3334/36/2/231

Ravindran, S., Hopkins, B., Telang, R., & Tingle, M. (2008). Metabolism and target characterization of a rat-selective toxicant norbormide. *Toxicology Letters, 180,* S198—S199. https://doi.org/10.1016/j.toxlet.2008.06.223

Regazzi, F. M., Justo, B. M., Vidal, A. B. G., Brito, M. M., Angrimani, D. S. R., Abreu, R. A., Lúcio, C. F., Fernandes, C. B., & Vannucchi, C. I. (2022). Prenatal or postnatal corticosteroids favor clinical, respiratory, metabolic outcomes and oxidative balance of preterm lambs corticotherapy for premature neonatal lambs. *Theriogenology, 182,* 129—137. https://doi.org/10.1016/j.theriogenology.2022.02.006

Santos, A. L. R., Wauben, L. S. G. L., Guilavogui, S., Brezet, J. C., Goossens, R., & Rosseel, P. M. J. (2016). Safety challenges of medical equipment in nurse anaesthetist training in Haiti. *Applied Ergonomics, 53,* 110—121. https://doi.org/10.1016/j.apergo.2015.06.011

Sarwar, A., Peters, R. T., Shafeeque, M., Mohamed, A., Arshad, A., Ullah, I., Saddique, N., Muzammil, M., & Aslam, R. A. (2021). Accurate measurement of wind drift and evaporation losses could improve water application efficiency of sprinkler irrigation systems — A comparison of measuring techniques. *Agricultural Water Management, 258*(October), 107209. https://doi.org/10.1016/j.agwat.2021.107209

Shaban, A., Abdelwahed, A., Di Gravio, G., Afefy, I. H., & Patriarca, R. (2022). A systems-theoretic hazard analysis for safety-critical medical gas pipeline and oxygen supply systems. *Journal of Loss Prevention in the Process Industries, 77*(March), 104782. https://doi.org/10.1016/j.jlp.2022.104782

Shaw, H. J., & Lin, C. K. (2021). Marine big data analysis of ships for the energy efficiency changes of the hull and maintenance evaluation based on the ISO 19030 standard. *Ocean Engineering, 232*(3), 108953. https://doi.org/10.1016/j.oceaneng.2021.108953

Shevchenko, V., Mialdun, A., Yasnou, V., Lyulin, Y. V., Ouerdane, H., & Shevtsova, V. (2021). Investigation of diffusive and optical properties of vapour-air mixtures: The benefits of interferometry. *Chemical Engineering Science, 233*, 116433. https://doi.org/10.1016/j.ces.2020.116433

Sloop, J. T., Donati, G. L., & Jones, B. T. (2022). Multi-internal standard calibration applied to inductively coupled plasma optical emission spectrometry. *Analytica Chimica Acta, 1190*, 339258. https://doi.org/10.1016/j.aca.2021.339258

Speers, A. J. H., Bhullar, N., Cosh, S., & Wootton, B. M. (2022). Correlates of therapist drift in psychological practice: A systematic review of therapist characteristics. *Clinical Psychology Review, 93*(February), 102132. https://doi.org/10.1016/j.cpr.2022.102132

Terrien, J., Marque, C., Gondry, J., Steingrimsdottir, T., & Karlsson, B. (2010). Uterine electromyogram database and processing function interface: An open standard analysis platform for electrohysterogram signals. *Computers in Biology and Medicine, 40*(2), 223–230. https://doi.org/10.1016/j.compbiomed.2009.11.019

Thompson, J. E., & Jaffe, M. B. (2005). Capnographic waveforms in the mechanically ventilated patient. *Respiratory Care, 50*(1), 100–108.

Wilson, M. (2021). Common errors in clinical measurement. *Anaesthesia and Intensive Care Medicine, 22*(3), 197–201. https://doi.org/10.1016/j.mpaic.2021.01.001

Winsen, Z. (2016). *NDIR Infrared CO_2 Gas Sensor - MH-410D*. https://www.isweek.com/product/ndir-infrared-co2-gas-sensor-mh-410d_143.html.

Wu, Y., & Li, W. (2010). Study on technology of electromagnetic radiation of sensitive index to forecast the coal and gas hazards. *Procedia Engineering, 7*, 327–334. https://doi.org/10.1016/j.proeng.2010.11.052

Wu, T. Y., Murashima, Y., Sakurai, H., & Iida, K. (2022). A bilateral comparison of particle number concentration standards via calibration of an optical particle counter for number concentration up to ~ 1000 cm^{-3}. *Measurement: Journal of the International Measurement Confederation, 189*(October 2021), 110446. https://doi.org/10.1016/j.measurement.2021.110446

Yaron, M., Padyk, P., Hutsinpiller, M., & Cairns, C. B. (1996). Utility of the expiratory capnogram in the assessment of bronchospasm. *Annals of Emergency Medicine, 28*(4), 403–407. https://doi.org/10.1016/S0196-0644(96)70005-7

Young, B., & Schmid, J. J. (2018). Updates to IEC/AAMI ECG standards, a new hybrid standard. *Journal of Electrocardiology, 51*(6), S103–S105. https://doi.org/10.1016/j.jelectrocard.2018.06.017

Zaharudin, S. Z. B., Kazemi, M., & Malarvili, M. B. (2014). Designing a respiratory CO_2 measurement device for home monitoring of asthma severity. *IECBES 2014, Conference Proceedings - 2014 IEEE Conference on Biomedical Engineering and Sciences: December*, 230–234. https://doi.org/10.1109/IECBES.2014.7047492.

Zhen, B., & Xu, J. (2013). Influence of the time delay of signal transmission on synchronization conditions in drive-response systems. *Theoretical and Applied Mechanics Letters, 3*(6), 063004. https://doi.org/10.1063/2.1306304

Zhevnenko, D., Meshchaninov, F., Kozhevnikov, V., Shamin, E., Belov, A., Gerasimova, S., Guseinov, D., Mikhaylov, A., & Gornev, E. (2021). Simulation of memristor switching time series in response to spike-like signal. *Chaos, Solitons and Fractals, 142*, 110382. https://doi.org/10.1016/j.chaos.2020.110382

Index

Note: Page numbers followed by "f" indicate figures and "t" indicate tables.

Printed in the United States
by Baker & Taylor Publisher Services